Armaments and Disarmament in the Nuclear Age

SIPRI
Stockholm International Peace Research Institute

SIPRI is an independent institute for research into problems of peace and conflict, with particular attention to the problems of disarmament and arms regulation. It was established in 1966 to commemorate Sweden's 150 years of unbroken peace.

The financing is provided by the Swedish Parliament. The staff, the Governing Board and the Scientific Council are international. As a consultative body, the Scientific Council is not responsible for the views expressed in the publications of the Institute.

SIPRI
Stockholm International Peace Research Institute

Sveavägen 166, S-113 46 Stockholm, Sweden
Cable: Peaceresearch, Stockholm
Telephone: 08-15 09 40

Armaments and Disarmament in the Nuclear Age

A HANDBOOK

SIPRI
Stockholm International Peace Research Institute

Humanities Press Inc.
Atlantic Highlands, N.J.,
U.S.A.

Almqvist & Wiksell
International
Stockholm, Sweden

Dr Marek Thee is a research fellow at the International Peace Research Institute, Oslo, and editor of the *Bulletin of Peace Proposals*.

First published by Almqvist & Wiksell International
26 Gamla Brogatan, S-111 20 Stockholm
(ISBN 91-85114-33-2)

in collaboration with

Humanities Press Inc,
171 First Avenue, Atlantic Highlands, N.J. 07716, USA

Library of Congress Cataloging in Publication Data
Stockholm International Peace Research Institute
Armaments and Disarmament in the Nuclear Age
"Edited by Dr. Marek Thee (i.e. M. Gdański)"
Bibliography: p. 299
Includes index.
1. Atomic weapons and disarmament.
2. Munitions.
3. War. (International Law). i. Gdański Marek. ii. Title.
JX1974.7.S762 1976 327.174 76-27299
ISBN 0-391-00652-5

Printed in England by
Taylor & Francis (Printers) Ltd, Basingstoke
and typeset by
The Lancashire Typesetting Co Ltd, Bolton
1976

Preface

The Stockholm International Peace Research Institute, SIPRI, was set up in 1966 as an independent institute for research into problems of peace and conflict. The policy adopted by SIPRI's Governing Board has been that the Institute should study problems in an independent and pragmatic way. It should select questions which are important to decision-makers in current international politics.

The Institute has so far mainly concentrated its research on problems on armaments, disarmament and arms regulation. The aim is to describe, as factually as possible, the major quantitative and qualitative changes that are taking place in the world's arsenals, and to analyse the efforts made to control these arsenals.

Over the past ten years SIPRI has produced a number of books and studies. The data is drawn only from open sources. The underlying value governing SIPRI's work is the belief that the world is devoting an excessive quantity of resources to armaments, and that this quantity could be reduced for human betterment, particularly with advantage to the underdeveloped countries.

To mark the tenth anniversary of SIPRI this compendium of work of SIPRI staff and visiting scholars on armaments and disarmament has been prepared. SIPRI publications are a collective effort. The credit for this book goes, therefore, to the entire staff, past and present, and to all the guest scholars who have worked at SIPRI over the past decade. The book contains shortened versions of published studies. References, valuations, tables, charts, registers, maps, texts of arms control agreements, detailed discussion and chronology can be found in *SIPRI Yearbooks* and other relevant SIPRI publications.

The book was edited by Dr Marek Thee while a visiting scholar at SIPRI in 1976.

FRANK BARNABY
DIRECTOR

OCTOBER 1976

Contents

TABLES AND CHARTS

Chapter 1. State of armaments

Table

Appendix 1A–D. World military expenditure, 1954–1975

Tables

Chapter 2. The nuclear momentum

Tables

Chapter 5. Conventional weapons and arms trade

Tables

Chapter 6. Armament dynamics and military research and development

Table

Chapter 7. Economic and social consequences of armaments

Tables

Chapter 8. Arms control and disarmament

Table

Charts

1. State of armaments

1. *Armaments expenditures, 1945–1975**

Total world military expenditure in 1975 was estimated to be $280 billion at current prices. Weapon production was large and widespread, the international trade in arms was extremely brisk and advances in military technology continued to frustrate efforts at arms control.

The post-war period is remarkable for the consistency with which large quantities of resources have annually been set aside for military purposes. Including 1975, cumulative world military expenditure since the end of World War II amounts to something like $4 500 billion. This figure is computed at constant (1970) prices and is almost certainly a conservative estimate. On the average, world military expenditure (in real terms) increased at an annual rate of 4.5 per cent between 1948 and 1975.

Although world military expenditure has moved upwards in spasms followed by periods of stability, the distribution of this expenditure has been changing in a more systematic way. The basic trend has been a declining concentration of expenditure on the European and North American continents with offsetting increases in the third world and China. The arms race has become a global phenomenon.

But despite the considerable and highly significant shift in the distribution of world military expenditure the NATO and WTO contributions remain predominant. NATO and the WTO totally dominate the world military scene. Four countries—the United States, the Soviet Union, the United Kingdom and France—provide the bulk of the world's capacity to design and produce weapons and, relatedly, virtually monopolize the international trade in arms, particularly with the third world. The combined military expenditures of the two alliances account for around 80 per cent of the world total. The United States and the Soviet Union are in a class of their own and essentially determine the military-technological environment for the rest of the world. Other countries accommodate themselves to this environment as their financial and technical resources permit.

Some comparisons

Total world military expenditure is (*a*) equivalent to the combined gross national products of the 65 countries in Latin America and Africa, (*b*) equivalent to total worldwide government expenditures on education, (*c*) about

* From *SIPRI Yearbook*, 1968/69, 1974, 1975 and 1976.

twice as large as government expenditures on health or (*d*) about 15 times as large as the value of all official assistance provided to the underdeveloped countries.

Comparisons such as these vividly illustrate the distorted priorities which have prevailed over the post-war period.

In the period since the Second World War, the world has given over to military uses much more of its output than it did either before the First World War or in the inter-war period. The quantity of resources devoted annually to armaments has, on the average, been more than five times as large since World War II than over the period 1925–38, or 7.5 times as large if the rapid rearmament immediately preceding World War II is excluded. In 1913, even after three years of a competitive arms race among the big powers, probably no more than 3–3½ per cent of world output was going to the military. In the early 1930's, the percentage seems to have been about the same. The average over 1950–70, on the other hand, has been around 7–8 per cent—more than double the 1913 figure. It now is over 6 per cent of the gross national product of the countries of the world and equal to the total income of countries whose populations comprise more than half of mankind.

Table 1.1. World military expenditure[a] *US $ bn, at constant (1970) prices*

Year		*Year*	
1908	9.0	1954	126.7
1913	14.5	1955	127.4
1925	(19.3)	1956	126.5
1926	(19.6)	1957	128.8
1927	(21.5)	1958	126.8
1928	21.5	1959	131.7
1929	21.7	1960	130.8
1930	23.2	1961	143.7
1931	21.9	1962	157.6
1932	20.3	1963	164.1
1933	20.1	1964	162.2
1934	23.9	1965	162.2
1935	32.6	1966	178.6
1936	47.1	1967	196.9
1937	58.8	1968	209.2
1938	61.6	1969	212.9
1948	64.7	1970	209.0
1949	67.9	1971	208.2
1950	73.5	1972	211.7
1951	107.0	1973	212.3
1952	137.2	1974	213.2
1953	140.9	1975	213.8

[a] Gaps in the chart are explained as follows: Before World War I, figures exist only for 1908 and 1913. After World War I, reasonably accurate figures are available for 1928 and onwards. Figures for 1925–27 can be adequately estimated (figures in parentheses). The post-World War II series begins in 1948 because expenditure in the first two post-war years was dominated by wartime levels of forces.
Source: SIPRI worksheets.

A recipe for an infinite arms race

When we combine this increase in the share of world resources going to military spending with the increase in world output itself, the result is a formidable rise over the last fifty years in the quantum of resources devoted to military uses. The world's national product has risen at least five-fold in the last fifty years: military spending, in real terms, has probably risen ten-fold. The reason is not so much that the world's standing armies are bigger —though they are bigger than they were fifty years ago. It is rather the immense increase in the cost and complexity of the weapons used.

The official view in many countries, instead of being one of concern, takes comfort in the mere fact that the ratio of military expenditure to GDP is generally falling, certainly in most of the industrialized countries. Such complacency is certainly unwarranted, however. Generally speaking this favourable trend in the allocation of resources is the result of relative increases in gross domestic product and non-military government expenditure; it is not the result of any reduction in military expenditure. If military spending maintains a constant share of world national output, this is a recipe for an infinite arms race. If things do in fact go on like this, then military spending will continue to double every fifteen years. By the early years of the next century the world will be devoting to military uses a quantum of resources which is equal to the whole world's present output. This is not so preposterous as it sounds. The world is now devoting to military purposes an amount of resources which exceeds the world's total output in the year 1900.

In this connection the enormous resources devoted to military research and development (R&D) should be mentioned. SIPRI estimates that about $20 billion per year is being spent on military R&D, or about one-third of the entire world expenditure on all R&D. The use of these resources for appropriate research and development for peaceful purposes could have an enormous effect in contributing to progress and development in the underdeveloped parts of the world.

The development and production of major weapons

The activity which lies at the heart of the world's arms race is the continuous procurement of new weapons. The weapons development and production programmes in the industrialized countries showed remarkable resilience to the economic difficulties which continued to plague most of these countries during 1975. Although economic recession and high rates of inflation produced an unfavourable climate, military budgets were maintained in most countries; some even increased their expenditures in real terms. And for those with competitive weapons to offer, the export markets, particularly in the third world, provided a volume of business more than sufficient to offset any stagnation in national demand.

At the end of World War II, only five countries—the United States, the Soviet Union, the United Kingdom, Canada and Sweden—had any significant capacity to develop major weapons. In 1973 some 30 countries were engaged in this activity, and others were manufacturing weapons of foreign design under licence.

The distribution of indigenous weapon development and production efforts is very uneven. Sizeable nuclear weapon programmes are carried out only in the United States and the Soviet Union, with much smaller efforts also under way in France and China. In the case of conventional (non-nuclear) weapons, the number of countries with an indigenous development capacity is much larger, but the volume of work remains concentrated in the four main arms-producing countries—the USA, the USSR, the UK and France. The US programme is the largest, and also, in most cases, involves the most advanced technology. The Soviet programme is also comprehensive, but it appears to be characterized by long production runs of basic designs, with the result that fewer different systems are in production at any one time. Together, the United Kingdom and France are producing about as many different major conventional weapon systems as all of the remaining developed countries combined.

Although the programmes of countries other than the four main arms producers are comparatively small, the volume of work in progress was very much greater in 1975 than it was 15 or even 10 years ago. In a number of developed and underdeveloped countries—including China, FR Germany, India, Israel and Japan—major long-term expansions in weapon development and production capacity are under way.

Horizontal proliferation of weapon production

Defence industries in third world countries continue to mature. Argentina, India and Israel have indigenous combat aircraft programmes under way; Brazil, India, Israel, South Africa and Taiwan are developing and/or producing their own missiles of various types—antitank, ship-to-ship, surface-to-surface, air-to-surface, and air-to-air. During 1975 Israel introduced its Kfir fighter, the first country other than the USA, the USSR, France, the UK, Sweden and China to develop successfully and produce indigenously—at least to a significant extent—an advanced, supersonic combat aircraft. Development and production activity is more widespread in less sophisticated fields such as trainer and transport aircraft, small ships and boats and small arms.

The horizontal proliferation of defence and defence-related industrial facilities also continued in 1975 with the establishment of aircraft-manufacturing plants in Peru, the Philippines and the Democratic People's Republic of Korea. However, the most important new development along these lines is likely to occur in Egypt which has a $2-billion fund—provided mainly by Saudi Arabia and Kuwait—for the establishment of an Arab defence industry, specifically the Arab States Military Industrial Organization.

The arms race in conventional weapons, and particularly the acquisition by more and more countries of the ability to develop and manufacture their own major weapons, is a neglected phenomenon, despite the fact that it is these weapons that have been used in all the wars fought since World War II. For this reason it is worth pointing out that the financial and technical resources devoted to the development and production of conventional weapons are far greater than those absorbed by the nuclear programmes.

Unless some positive steps are made toward global disarmament, the horizontal and vertical proliferation of conventional weapon production capacity will almost certainly continue.

2. *Nuclear megatonnage**

Estimates of total megatonnage are few and need to be interpreted with care.

In March 1960 Senator (later President) J. F. Kennedy stated that the world's nuclear stockpiles amounted to about 10 tons of TNT for every person on the globe. This means a total of about 30 000 megatons. Professor York states that the 1960 Pugwash meeting used 60 000 megatons as a working assumption, and that 'we are therefore safe in assuming that the United States possessed at the beginning of the sixties a strategic weapon stockpile containing 20 000 to 40 000 megatons of explosives'.

Professor Pauling estimated the total megatonnage in the possession of the Soviet Union and United States at 250 000 megatons in 1962 and 320 000 megatons in 1968, but he appears to have arrived at his figures by assuming that most or all of the fissile materials would be made up into thermonuclear weapons of high yield. His figures thus appear to indicate what might theoretically be produced rather than what was produced. Senator Kennedy's figure and Professor Pauling's figure are estimates of the megatonnage of all warheads, including those for reloads, reserves, surpluses and so on. If the total stands at 50 000 megatons, which seems a reasonable guess, it would represent about 15 tons of TNT per person on the globe, or—and this is more meaningful—about 60 tons per person in the NATO and Warsaw Pact nations taken together. Such an 'overkill' is so fantastic that whether the true figures is twice or half as high seems to matter little.

There have been various statements about what megatonnage could be delivered in a single attack by the forces of different nuclear powers. For the United States an estimate in 1961 was that the Strategic and Tactical Air Commands could lift 18 000–20 000 megatons against the Soviet Union in 24 hours. Other estimates, both higher and lower than this, were made in the United States in the early 1960s; but it is not always clear what they mean. Some refer to 'alert forces' only, a term which may not have a constant meaning over time, as weapons systems change. Others refer to the total number of warheads in delivery systems whether 'on alert' or not, but it is not always clear whether in fact all delivery systems are included. Some delivery systems, 'tactical' or otherwise, usually seem to be excluded.

* From *SIPRI Yearbook*, 1969/70.

*3. Number of nuclear warheads**

It is important, to begin with, to recognize the variety of weapons for which nuclear warheads have been fabricated. Reference has been found to nuclear weapons in the following categories for the United States.

1. ICBMs
2. MRBMs and IRBMs
3. free-fall bombs
4. air-to-surface standoff missiles
5. air-breathing cruise missiles
6. surface-to-air missiles
7. air-to-ground missiles
8. air-to-air missiles
9. army and naval artillery
10. depth charges
11. torpedoes and rocket torpedoes
12. ocean mines
13. atomic demolition devices or land mines

The size of warheads varies very widely. Artillery shells may have an explosive charge of 2 kilotons or so. The largest warheads appear to be around 25 megatons—1 250 times larger. Generally speaking, the trend is to more and smaller warheads, as accuracies have increased and the possibility of multiple warheads has been developed. This is now the way to maximize damage.

The United States stock of warheads has often been estimated by dividing the estimates of the stock of fissile material by estimates of the amount required to make a 20-kiloton fission bomb. This procedure may not be too bad for estimating numbers of warheads (as opposed to megatonnage), though even then the estimate must be regarded as being in terms of 'nominal' warheads rather than the actual mix of warheads in stock. On this basis different experts have estimated that the number of US warheads was variously 100 000, 150 000 and 200 000, in the late 1960's. Disagreement to this extent is not surprising.

Indications of the number of real warheads were given in various official statements in the United States in the early 1960's. Mr Gilpatric, Assistant Secretary of Defense, stated that:

> The total number of our nuclear delivery vehicles, tactical as well as strategic, is in the tens of thousands, and of course we have more than one warhead for each vehicle.
> The Arms Control and Disarmament Agency put the total number of American warheads at 40 000 in 1962 . . .

Mr McNamara referred to 'tens of thousands of nuclear explosives for tactical use'. Since the tonnage of fissile material appears to have increased

* From *SIPRI Yearbook*, 1969/70.

two- or three-fold since that time, there seems nothing implausible in the idea that the number of United States warheads may be of the order of 100 000.

At various dates there have been statements about how many warheads were in the 'alert force': 850 in 1961; a 100 per cent increase in three years ending some time around 1964; 2 200 and 2 600 in 1967. Exactly which weapon systems comprise the 'alert force' and how this group is defined are not clear.

There have also been statements about the number of US warheads in Europe—several thousand in 1963, 7 000 and 7 200 in 1968.

The only estimates available for the Soviet Union come from Western sources. These are often made by stating that the Soviet stock of warheads is believed to be a certain per cent of the United States stock. The percentages given in the different sources are inconsistent with one another: 3 per cent in 1953, 10 to 20 per cent in the early 1960's, as high as 60 per cent in 1961, and back to less than 10 per cent in 1966. The available estimates of absolute numbers are 100 to 300 warheads in 1953, 5 000–10 000 in 1964, 5 000–10 000 in 1966.

Little is known about other countries. In 1964 it was estimated that Britain had 'perhaps 1 500' warheads. No estimate of warhead numbers is available for France, but judging from the number of delivery vehicles the figure may run into the hundreds. It is estimated in this year's US posture statement that in China 'the amount of U 235 now estimated to be available for stockpiling would be sufficient for only a few dozen weapons of any type'.

4. *Strategic nuclear weapons**

The strategic weapon programmes which were underway in the United States and the Soviet Union at the conclusion of the first round of the Strategic Arms Limitation Talks (SALT) in May 1972, and new initiatives taken by both sides since that time, have now brought the two countries to a critical juncture in the strategic arms race. New generations of strategic weapons loom close and the systems that may follow these are already beginning to take shape in the plans of the weapon designers and the arguments of the strategists.

It is generally held that the main trend in US and Soviet strategic weapon developments is toward qualitative improvements rather than quantitative increases. The number of offensive nuclear-weapon delivery vehicles—bombers, intercontinental ballistic missiles (ICBMs), and submarine-launched ballistic missiles (SLBMs)—has tended to level off and, rather than deploying larger numbers of these aircraft and missiles, the USA and the USSR are replacing existing systems with newer and more effective ones.

* From *SIPRI Yearbook*, 1974.

Quantitative expansions

Despite the tendency of the number of strategic delivery vehicles on the two sides to level out, significant quantitative expansions in nuclear forces are continuing. The most important of these is, of course, that which results from MIRVed (multiple independently targetable re-entry vehicle) missile warheads, which are currently being deployed by the United States and developed by the Soviet Union. The US deployments involve the replacement of the majority of existing land- and submarine-based missiles by new MIRVed missiles (Minuteman III and Poseidon) in programmes which, between 1970 and 1977, will produce a five-fold increase (from about 2 000 to about 10 000) in the number of independent nuclear warheads that can be delivered by the missile forces.

The development of MIRVs has given the United States a very large quantitative lead over the Soviet Union in the number of deliverable nuclear warheads. However, since the Soviet Union is currently developing MIRV technology and has deployed a larger number of land-based missiles, and missiles with a greater 'throw-weight', it has the potential to close the gap and possibly exceed the USA in numbers of warheads.

It is surprising that in the general assessment of the results of SALT I, the US advantage in MIRV technology was often cited as the main *qualitative* advantage on the US side offsetting the quantitative advantage in ICBMs and SLBMs permitted to the Soviet Union under the Interim Agreement on offensive strategic weapons. While the US lead of five years or more in the development of MIRV systems is undoubtedly a good indicator of a more general US technological advantage, the significance of the MIRV lead lies mainly in its quantitative impact. There are other areas of technology in which the United States has long held qualitative advantages which cannot easily be incorporated in an overall quantitative comparison of the forces. These include five important areas in which continuing advances have been made by the United States in the period since SALT I.

Race in quality

The first is missile accuracy, which provides the capability to destroy ICBMs in hardened silos. The Minuteman III and Poseidon missiles currently being introduced into the US forces have an accuracy of about one-quarter of a nautical mile, an improvement over the previous land- and sea-based missiles by a factor of two. Current Soviet ICBMs and SLBMs are generally credited with an accuracy of about one mile. Programmes to improve accuracies are underway in both the United States and the Soviet Union: but whereas improvements on the US side are clearly within reach, given the present state of US technology, the latest systems under development in the Soviet Union are reported not to have shown any significant improvement in accuracy.

Second, all 1 000 modern US ICBM silos are 'hardened' (reinforced with structures of concrete and steel) to withstand nuclear blast overpressure of about 300 pounds per square inch (psi). These silos are now being upgraded to a level of at least 900 psi and possibly considerably more. In the case of the Soviet Union, only two-thirds of the current ICBM force is believed to be emplaced in silos capable of withstanding 300 psi overpressure: silos of the other missiles, including the large 'SS-9s', are estimated to have a 100 psi resistance or, in the case of the older 'SS-7s' and 'SS-8s', as little as 5 psi. The 90 latest Soviet silos are believed to have been hardened to 600 psi but there is no evidence of substantially increased hardening of the 1 527 earlier missile silos.

Third, a new US advantage in ICBMs, introduced within the past year, is that of 'remote retargeting' of missiles from launcher control facilities. Even earlier, the United States had some lead in this area, since it could 'preprogramme' Minuteman II missiles with up to eight alternative targets, as compared with the one to two targets which could be set in the earliest US ICBMs or in present Soviet missiles. For the Minuteman III, the number of alternative targets is essentially unlimited. This capability is useful in the event of a US counterforce attack against Soviet ICBMs, since it permits rapid and flexible replacement of first-round missiles which are observed to fail.

Fourth, continuing advances have been made by the United States in strategic bomber range and payload (including advances in the B-1 bomber, which is still under development). There have also been improvements in 'escape time' and in the resistance of installed equipment to the effects of electromagnetic pulses, which increase the survivability of the bomber fleet in the event of an attack by Soviet nuclear forces.

Finally, US naval officials estimate that the United States has a considerable lead in the quietness and reliability of its strategic submarines. This increases their invulnerability to antisubmarine warfare (ASW) efforts. Installment of 'submarine quieting' equipment which will further improve the performance of the US strategic submarines is being undertaken at the same time as the fitting of the new MIRVed missile (Poseidon).

Submarine-launched ballistic missiles

One new area of qualitative advantage on the Soviet side has been observed in the past two years. At the end of 1972 the Soviet Union tested a new SLBM ('SS-N-8') out to a range of 4 200 nautical miles—a vast improvement over its first longer-range SLBMs (the 'SS-N-6', with a range of 1 500 nautical miles, equaling that of the US SLBMs introduced in 1962) and one which gave the Soviet Union a considerable advantage over the United States in SLBM range. However, when payload is taken into account, the Soviet advantage declines. The 'SS-N-8' is believed to carry about the same payload as the US Polaris A-3; and the current US Poseidon C-3, which has the same range as the Polaris A-3 (2 500 nautical miles), is reported to weigh three times as much as, and to carry twice the payload of, the A-3. A reduction of the larger payload of the Poseidon would result in increased range: the potential range of the Poseidon has been kept secret, but published estimates range from more than 3 000 nautical miles to 4 300 nautical miles.

Improvement in counterforce

The main effect of the on-going US programmes in offensive strategic weapons is to increase the invulnerability of the forces to Soviet attack or countermeasures and to improve their capability to attack Soviet land-based nuclear forces. This improvement in counterforce capability results mainly from the continuing increases in the number of missile warheads and in the accuracy of the delivery of these warheads—two areas in which the United States has a great advantage over the Soviet Union. The silo-hardening and 'submarine-quieting' programmes, along with the development of a new, longer-range SLBM (Trident), account for the main increase in the invulnerability of the forces.

Unlike the US programmes, recent Soviet deployments have done little to increase the counterforce capabilities of Soviet strategic forces. However, the deployment of the new, longer-range Soviet SLBM will provide the USSR with a very considerable increase in the invulnerability of its submarine-based force to US ASW. With the longer-range missiles, the Soviet strategic submarines will be able to cover US targets without traversing areas where they are more subject to detection.

Soviet testing of new, large MIRVed ICBMs, with a relatively large number of warheads (four to six) suggests a potential to develop a land-based missile force with counterforce capabilities comparable with those presently being introduced in the United States. The exploitation of this potential would require the development of greatly improved missile accuracies, as well as full use of the large potential throw-weight of the new missiles. Over the long term, a capability to destroy virtually all of the US ICBM force might be evolved. The Soviet lag in missile accuracy is, however, such that it would probably require a considerable time to develop this potential, particularly in view of the super-hardened silos under construction in the United States. The US programmes to improve ICBM guidance, warhead yield and warhead numbers would permit a much more rapid US acquisition of the capability to destroy the entire Soviet land-based missile force.

5. *Tactical nuclear weapons**

In 1944, no nuclear weapons existed. Today, there are tens of thousands. The nuclear arsenals of the USA and the USSR have grown so large as to

* From *Disarmament or Destruction*, SIPRI, 1975.

be wildly in excess of any conceivable need, military or political, of either power.

Thirty years ago, nuclear weapons could be delivered only by bombers. Nowadays, they exist in a bewildering variety of forms: intercontinental ballistic missiles (ICBMs), submarine-launched ballistic missiles (SLBMs), medium-range ballistic missiles, intermediate-range ballistic missiles, short-range ballistic missiles, depressed trajectory ballistic missiles, fractional orbital bombardment systems, free-fall tactical bombs, free-fall strategic bombs, air-to-surface missiles, air-to-surface stand-off missiles, air-to-air missiles, army artillery shells, naval artillery shells, howitzer projectiles, torpedoes, rocket torpedoes, depth charges, demolition devices, land mines, sea mines, anti-ballistic missiles and so on.

50 000 Hiroshimas

Most publicity has been given to strategic nuclear weapons, but it should not be forgotten that in addition to huge strategic forces, the USA and the USSR have deployed tens of thousands of tactical nuclear weapons. In Europe alone there are about 7 000 US and 3 500 Soviet tactical nuclear weapons. The US weapons are widespread—in the Federal Republic of Germany, the Netherlands, Belgium, Italy, Iceland, Spain, Portugal, Turkey, Greece and the UK. In addition, the USA, for example, has deployed nearly 2 000 tactical nuclear weapons in Asia, mainly in Korea and the Philippines but also in US bases in Guam and Midway. There are thousands of nuclear weapons at sea. The US Navy, for example, has at least 2 500, and possibly many more, nuclear weapons deployed in the Atlantic and Pacific Fleets. And then there are all of the nuclear weapons deployed or stored in the two big powers themselves. It is estimated that the USA keeps at least 10 000 tactical nuclear weapons on its own territory. The USA has, therefore, deployed or stockpiled worldwide, a total of considerably more than 20 000 tactical nuclear weapons. And the Soviet Union has probably manufactured a similar number of these weapons—we do not know for certain because, regrettably, the USSR does not publicise this type of information.

The total explosive power of existing tactical nuclear weapons is enormous—so large as almost to defy imagination. The combined explosive capability of US and Soviet tactical nuclear weapons is roughly equivalent to 700 million tons of TNT or 50 000 Hiroshimas. The tactical nuclear weapons at present deployed in Europe would, if ever used, totally obliterate the continent—20 000 Hiroshimas would be unleashed.

6. *Nuclear miniweapons**

Miniweapons, a new generation of tactical nuclear weapons, are presently being introduced in Europe. They have smaller yields than the

* From *Nuclear Proliferation Problems*, SIPRI, 1974.

earlier types (possibly as low as a few tons of TNT equivalent), produce little blast and almost no radioactive residue. Being laser-guided they are highly accurate. Nuclear shells of 155-mm and 8-inch howitzers, and possibly artillery rockets and atomic demolition munitions (ADMs) are being miniaturized. The main drawback of such modernization is an increased probability of the real use of nuclear weapons in case of crisis. In addition, this modernization is contrary to the spirit of the NPT.

The first indication that such weapons were actually being developed came from former US Secretary of Defense Melvin Laird in an interview on 13 April 1972. Mr Laird announced that the US government had been developing and improving its tactical nuclear weapons for a number of years. The main purpose of this modernization programme was to obtain smaller yields and less radioactive residue as well as improved accuracy, flexibility and safety. Mr Laird stressed in the interview that a successful conclusion of SALT would result in greater emphasis being placed upon the role of tactical nuclear weapons. Advocates of modernization maintain also that enhanced radiation weapons would be sufficiently different, in a qualitative sense, to permit the establishment of a firebreak between them and present-day tactical nuclear weapons, that is, that they could be used like conventional weapons without the risk of escalation.

Intended for real use

Low-yield howitzer shells provided with a homing device do not necessarily represent the most typical miniweapons. By nuclear miniweapons we usually understand low-weight and low-yield one-man devices that can be carried in a handbag and launched by a mortar or bazooka, or dug into a pit as an atomic demolition munition (ADM) and discharged from a distance. They would be developed for every conceivable purpose and deployed in large numbers by the front-line units and commandoes. They would be intended for real use, and their use would probably be unavoidable in the case of war.

Such typical miniweapons are feasible and evidently have been developed, but it is not yet known to what extent they have been deployed.

Their deployment is a double-edged sword. They would certainly be useful and decrease the amount of collateral damage, which is good, of course. But the trouble is that their usefulness would simultaneously guarantee their deployment in case of war. This, again, would almost certainly trigger a process of escalation, which would lead to a full-scale nuclear war. There are several inbuilt escalation mechanisms in the use of nuclear weapons which make the restriction of their use extremely difficult. Because of the effectiveness and vulnerability of these weapons, both sides are under tremendous pressure to use their own nuclear weapons for destroying those of the adversary before he can use them. This may lead to large-scale pre-emptive strikes. Use of nuclear weapons soon destroys target acquisition and communication possibilities. Radio and radar connections may be cut for hours or days. This leads to escalation of yields, since lower accuracy must be compensated by higher yields. Furthermore, one is always more liberal regarding the weapon yields intended against the enemy than those which one has to absorb oneself. Therefore, the two sides can be expected to increase the yields alternately in each salvo. Finally, the bitterness of war leads to the desire to punish and destroy beyond any strategic or tactical need, as shown by the terror bombardment of Dresden in World War II.

The development of a new generation of nuclear weapons, the mini-weapons, shows that the nuclear powers have no intention of giving up their nuclear weapons.

7. *Trends in the nuclear arms race*

Developments in the first two decades after the Hiroshima bomb were basically a race in quantities, moving from the atom to the hydrogen bomb, from lower to higher yields in single warheads. The subsequent stages of this race were made visible by the escalation of yields from the largest World War II pre-nuclear bomb, the 'blockbuster', with an explosive power of *ten* tons of TNT, to the Hiroshima bomb with a yield of fifteen *thousand* tons of TNT, to warheads now counted in *millions* of tons of TNT. It was thus an upward race from kiloton to megaton bombs, with a simultaneous feverish quantitative build-up of nuclear arsenals.

In the process, the nuclear powers acquired a potential for massive destruction unparalleled in history. The very order of magnitude has radically changed. As a measure of comparison, we may recall that all the bombs dropped by the United States in World War II on Germany and Japan had a cumulative explosive power of two megatons of TNT, while today one larger nuclear warhead in the arsenals of the United States and the Soviet Union may surpass this figure 10–15 times. Taken together, the nuclear warheads now in the hands of the great powers have an explosive power exceeding a million times the yield of the Hiroshima bomb, enough to wipe out humanity many times over.

Parallel and following the development of the bomb, efforts were made to advance the technology of delivery vehicles. The great strategic break-throughs came with the introduction, around the time of the first Sputnik in 1957, of intercontinental ballistic missiles, the ICBM. In quick succession followed then the low-penetration bomber; the forward land-based missiles, the supersonic bomber, the heavy ICBM, and the high-speed re-entry missiles.

From quantity to quality

All the above developments accentuated the threat policy of massive destruction. But the tools acquired were still low in manoeuvrability and operational use. It was only in the past decade that development of nuclear arms was channelled from quantity to quality moving dramatically ahead in improved usability and efficiency. Nuclear warfare was made thinkable. With advanced weapons at hand, military strategists could develop fanciful scenarios of 'limited' nuclear warfare, 'flexible' options and 'selected' targeting, 'exchange' of nuclear blows and low-collateral damage.

As far as nuclear warheads are concerned, attention turned to the marriage of their explosive power with greater technological efficiency, and to development of bombs with smaller charges but highly advanced performance. In hard target destruction, a doubling of accuracy proved to compensate for an eightfold reduction in yield. The race thus shifted to qualitative improvements: higher speed, and greater range, more precise targeting and guidance, greater penetrability and manoeuvrability, better performance and reliability, and automation.

In this context, the strategic nuclear arsenal was supplemented by a large variety of tactical nuclear weapons, their size ranging from some Hiroshima bombs to sub-kilo levels. They differ from strategical nuclear weapons mainly in range. Though with a shorter range they are sometimes as powerful as strategic weapons. It is estimated that actually all the nuclear powers accumulated a stockpile of nuclear warheads, strategical and tactical, growing towards the 100 000 mark—much in excess of the number of targets the most exalted military planner could invent. The last in the series of these weapons is the so-called miniature nuclear bomb or' mini-nuke' combining relatively low explosive power with extreme precision of targeting and possibly reduced radioactive fall-out. The mini-nuke is especially designed to blur the threshold between nuclear and conventional warfare so as to make it more feasible and palatable.

At the same time, the perfecting of the bomb was synchronized with the advancement of missile technology. The most important breakthroughs in this field were the on-board equipment of missiles with computers and the development of the MIRV—the Multiple Independently Targeted Re-entry Vessels, i.e. ballistic missiles with many warheads earmarked for different targets. The last word in this series is the MARV—Manoeuvrable Re-entry Vessels capable of changing direction in flight to evade possible defensive missiles.

In sum, from a stage of rough, uncertain and artless performance, nuclear warfare was lifted in the minds of military planners to highly sophisticated and operational levels. Nuclear weapons were streamlined and adapted for ready consumation. The push for modernization and innovation knows no pause.

8. *World stock of fighting vessels**

Over the period 1950–74, the estimated value of the world stock of fighting ships increased threefold, roughly the same as total world military

* From *SIPRI Yearbook*, 1975.

expenditure in constant prices. The rates of growth for different countries and regions have varied significantly. As a result there have been marked changes in the distribution of the world naval stock.

By far the most significant changes has been the emergence of the WTO—in essence the Soviet Union—as a major naval power. Over the period 1950–74, the WTO's naval stock has, on the average, increased twice as fast as the world total and three times as fast as that of NATO. The countries of these two alliances account for the lion's share of the world's military resources, and naval forces are no exception. The USA and the USSR dominate these respective alliances, each maintaining a navy many times larger than that of any other nation.

The USA and the USSR

Until comparatively recently, any joint discussion of the navies of these two countries would have been incongruous. Throughout the 1950s and early 1960s the US Navy was far larger than that of the Soviet Union, although the US advantage, in terms of the estimated value of fighting ships, declined from 6.5:1 in 1960 to 2:1 in 1965. Indeed it was not until 1960 that the Soviet Union replaced the UK as the country with the second largest fleet in the world (in terms of estimated value).

Thus, it is only in the past ten years or so that the Soviet Union has possessed a fleet large enough to be realistically compared with that of the United States. It seems worth pointing out here that the *speed* at which the USA has lost its status as the world's unrivalled naval power is probably in large part responsible for the many alarmist assessments made in recent years of the reasons for, and the consequences of, the expansion of the Soviet Union's naval forces.

Although the estimated capital values of the US and the Soviet navies were roughly equivalent in 1974, this equivalence disguises quite marked differences in the structure of their naval forces. The navies of the United States and the Soviet Union have at least one function in common—strategic deterrence. In both countries, nuclear-powered ballistic missile submarines constitute a major part of the strategic nuclear forces. And in both countries these vessels account for a large share of the estimated value of the naval stock, 28 and 36 per cent for the USA and the USSR respectively in 1974.

The significance of the Soviet lead in numbers (and value) of strategic submarines (48 against 41 in 1974) is minimal at the present time. If one accepts that the purpose of these boats (together with land-based missiles and bombers) is to deter nuclear attack, then both countries have more than enough already, so that a numerical inequality means very little except, it seems, in political terms. If the index used is the number of targets that the missiles in these submarines can threaten, then the advantage is sharply reversed with the USA having a lead of over 7:1 in 1974. By the end of 1974 about 26 of the 41 US boats carried the Poseidon missile with ten individually-targetable warheads while none of the Soviet submarine-launched missiles had this capability. On this basis, therefore, the US vessels should be assigned a much higher value than those of the Soviet Union. To estimate an appropriate value differential based on the number of deliverable warheads and other factors, such as missile accuracy and quietness of the submarines, would however, be a complex exercise even if the data were available. Moreover, it would not be particularly meaningful.

Apart from strategic deterrence, the principal elements of US naval strategy have remained fairly stable over the post-war period. After strategic deterrence, the greatest importance is attached to control of the sea. Experts argue at great length about the precise meaning of this phrase, but the following is probably an acceptable definition for the layman. Control of the sea means that one's own shipping—primarily merchant shipping—can use the oceans in time of war without incurring unacceptable losses and that this flexibility is denied the enemy. Next comes the projection of military power overseas, which essentially means the ability to bring military power to bear in distant countries. Even today the USA possesses by far the most powerful and sophisticated capability for this function with 15 attack aircraft carriers and a 200 000-man Marine Corps with its own specialized equipment for amphibious operations, particularly seven 18 000-ton amphibious assault ships. The final element is an overseas presence, that is, the ability to deploy naval vessels periodically or continuously in distant oceans. As mentioned above, the objectives of this activity are difficult to characterize, but it is nonetheless considered to yield substantial dividends.

A major naval build-up

There can be little doubt that a major naval build-up is under way around the world and will persist if present construction programmes maintain their momentum. For the foreseeable future, however, naval developments will largely depend on events in, and relations between, the USA and the USSR; in 1974 these two countries accounted for more than 70 per cent of the estimated value of the world naval stock.

There is every indication that the USA and the USSR are on the verge of a naval arms race if, indeed, it has not already begun. This study (and most others) concludes that the USA is still the world's strongest and certainly the most flexible naval power. But for the first time in the post-war period there is now a serious rival. Soviet naval strength has now reached the point at which it is possible to argue that further (unmatched) expansion will threaten US naval superiority. In other words, a necessary condition for a naval arms race, previously lacking, now exists. Moreover, the USA and other NATO powers have shown themselves to be acutely sensitive to the emergence of a rival naval power and so far the evidence suggests that they propose to accept the challenge. In the USA, the value of new naval construction under way or planned for the rest of this decade has been estimated at $21.4 billion and US Navy officials are confident that if these plans and those for the conversion and modernization of existing ships are carried through, a clear US naval superiority will be retained. However, it is only to be expected that the Soviet Union will match, in a general sense, any further US build-up.

To defuse the impending naval arms race

A naval arms race between these two countries that has the objective of securing some freedom to exploit their seapower for diplomatic and political ends would be highly dangerous. Because the USA and the USSR each possess nuclear arsenals capable of destroying the other many times over, and much of the rest of the world besides, the greatest stress has been placed on preventing any confrontation anywhere between the armed forces of these two countries. Unless immediate steps are taken to defuse the impending naval arms race between these two countries, such confrontations can be expected to occur with increasing frequency.

It is therefore urgent that negotiations begin between—to start with, at any rate—the USA and the USSR on the limitation of conventional naval armaments. There is, fortunately, a considerable historical precedent for such negotiations providing guidance both on how and how not to construct naval limitation agreements. The nuclear arms race has demonstrated the futility of the quest for superiority. Such superiority as is achieved is usually limited, temporary and unexploitable either for military or for political ends. The nuclear arms race has also demonstrated that the greater the proliferation, the harder it is to stop the process. The initiation of naval arms limitation negotiations would save both countries enormous expense and would forestall a contest that involves a high risk of confrontation.

9. *Stockpiles of chemical weapons**

With the exception of chemical irritants and herbicides, it appears that very few states today have modern chemical weapons at their disposal in any significant quantity. At least a dozen did so at the time of World War II, but over the years since then most of the stocks have either been destroyed or allowed to deteriorate in storage, probably beyond the point of military utility. The indications are that only certain of the nuclear-weapon states resumed production after World War II. It is conceivable, however, that, in view of allegations made during the past 25 years, certain other states may also have acquired chemical weapons, most notably in the Middle East; but, apart from the reports emanating from the Yemeni Civil War of the mid-1960's (and even these—the publicly disclosed ones, at least—do not offer complete verification), there is no reliable evidence either to confirm or refute this surmise.

* From *Chemical Disarmament: New Weapons for Old*, SIPRI, 1975.

Substantial quantities of modern chemical weapons undoubtedly exist in the NATO/Warsaw Treaty Organization (WTO) area. The United States makes no secret of the existence of its own stockpiles, but with the exception of France there is no evidence to suggest that other NATO countries still have useable supplies of their own. Part of the US stockpile is stored in the Federal Republic of Germany, but it is under US rather than NATO control. As regards the WTO countries, there have been no statements by Soviet or other state officials, or official publications in the open literature, about the existence or non-existence of an offensive CW capability. Nevertheless, NATO authorities appear to be convinced that the WTO does indeed possess such a capability. There have been numerous disclosures of Western intelligence appraisals of the matter which purport to give details of Soviet chemical weapons. These need to be treated with some caution, if only because they have tended to appear at times when Western CW programmes have come under popular attack or budgetary scrutiny.

It appears that the United States has about 35 000 tonnes of mustard and nerve gases available today, about 40 per cent of which are sarin (GB) and 10–15 per cent VX nerve gases. Somewhere in the region of half of the total agent stockpile has been filled into munitions, the remainder being held in bulk storage. The greater part of the filled-munition stockpile (which probably amounts in all to 150 000–200 000 tonnes), and probably all of the agent in bulk storage, is kept within the continental United States. Most of the nerve-gas munitions deployed in US depots in Europe are from the Army inventory of 105-mm, 155-mm and 203-mm artillery projectiles, 115-mm rockets, and landmines. Several other nerve-gas munitions have also been approved for the operational inventory, including warheads for Army tactical rockets and guided missiles, projectiles for Naval ordnance, and several types of bomb, cluster-bomb unit (CBU) and spray-tank, but, with the exception of certain of the aircraft spray-tanks and older designs of bomb, it has not been disclosed whether these have been procured in any substantial quantity.

The Soviet chemical stockpile is, according to a 1969 US official source, seven to ten times larger than the US one. Figures of 350 000 and 700 000 tonnes of CW agent have been quoted in the West German military press, but European NATO officials have expressed scepticism about them, regarding even the lower figure as a substantial exaggeration. The greater part of the stockpile is said to consist of mustard gas and other agents of World War I and II vintage, but reference has also been made to stocks of the nerve gases tuban, soman, and something of unknown identity called VR-55. There has been no serious mention of Soviet incapacitating agents. West German military journalists have claimed, six years ago, that 35 per cent of the Soviet stockpile of artillery munitions, bombs and rockets (25 per cent of the artillery munitions and 45 per cent of the rockets), and one-third of the Soviet wooden landmines stored in the German Democratic Republic, are loaded with CW agent. Western sources also assert that, among currently deployed Soviet weapon systems, chemical payloads are available for 122-mm and 152-mm artillery, the 122-mm multi-barrelled rocket launcher, T-5 series rockets (for example, what NATO calls the FROG-7), T-7 series tactical guided missiles (for example, SCUD-A), anti-personnel landmines, and certain unspecified aircraft munitions. It is said that Soviet chemical munitions have been made available to other WTO countries.

10. *Reconnaissance satellites**

One of the remarkable features about the SALT I agreements is the relative ease with which the problem of verifying the implementation of these agreements was solved. The requirement of on-site inspection, which has been considered by many states as essential in checking compliance with arms-control treaties, has been totally discarded in SALT I. For the first time the United States, the main proponent of on-site inspection, admitted that the modern means of verification at the disposal of the great powers are superior and more reliable than on-site inspection for monitoring the quantitative limitations of arms. On the other hand the Soviet Union, for the first time, accepted the principle of 'open skies', proposed by the United States as long ago as July 1955, but only as far as satellite altitudes are concerned. As a matter of fact it is public knowledge that both sides have, for a number of years, been using satellites for the purpose of intelligence gathering, although neither side has officially admitted it.

An obvious limitation of such control methods is their inability to check qualitative changes in the military arsenal of states, although development of certain new weapons should be detectable at the testing stage. But even for verification of quantitative limitation of arms, an obligation not to use concealment for impeding verification is important. This has been provided for in SALT I agreements, and would also have to be provided for in other arms-control agreements. However, the development of multi-spectral sensors carried by satellites may make it possible, in future, to detect certain objects under camouflage; for example, underground missile silos could be detected using infrared sensors to distinguish the heated silo from the surrounding.

In all such verification operations a prohibition on interference with satellites would be essential. The concept of verification by satellites could be jeopardized by the development of techniques whereby satellites can be intercepted and destroyed by other satellites.

Types of satellite

Until 1972, the United States had three types of photographic reconnaissance satellites: area-surveillance, 'close-look' and the large 'Big Bird'.

* From *SIPRI Yearbooks*, 1973, 1974, 1975 and 1976.

The first of these carried a wide-angle, low-resolution camera for searching a large area of a particular country for objects or events of potential interest. The 'close-look' satellite carries a high-resolution camera which has a very long focal length lens and a relatively narrow field of view. The 'Big Bird' satellite is designed to perform both the area-surveillance and the close-look type of mission. The Soviet Union has also area-surveillance and close-look satellites. Satellites which carry photographic equipment with high resolution are usually manoeuvrable.

Electronic reconnaissance satellites are launched into orbits with perigee heights of about 300–500 km and have considerably longer orbital lives. As the satellite passes over an area of interest, radar signals and other sources of electromagnetic radiation are recorded on tapes: the tapes can then be played back, and the recorded information transmitted to ground receiving stations.

Early-warning satellites use infrared techniques to detect the launch of enemy missiles. These satellites orbit at very high perigee heights, usually greater than 30 000 km, and have very long orbital lives, greater than one million years.

Ocean surveillance

The number of US reconnaissance satellites launched each year seems high, particularly if the satellites are used only for verifying the implementation of the SALT I agreements. It is believed that some of the older generation satellites which were launched soon after Big Bird satellites may be carrying infrared equipment for ocean-surveillance purposes. The infrared equipment carried by some of the US reconnaissance satellites is similar to that used in the Earth Resources Technology Satellite (ERTS 1). Photographs taken by cameras on ERTS 1 from a height of 900 km even show such details as small pleasure boats. The resolution would improve by a factor of six if photographs are taken from a height of about 150 km, the perigee height of most US photographic reconnaissance satellites. Such infrared devices are still in the development stages but, when they are fully developed, it may be possible to detect nuclear submarines travelling at considerable depths.

The Soviet Union has been using satellites to survey the oceans of the world since 1973. These satellites perform their ocean-surveillance missions in pairs: for example, Cosmos-651 and Cosmos 654 launched in 1974, and Cosmos-723 and Cosmos-724 launched in 1975.

An analysis of satellites launched during the past few years shows that since 1968 the Soviet Union has launched an average of about 30 photographic reconnaissance satellites per year whereas the rate for the United States has been about seven per year since 1971. The rate of satellite launches by the Soviet Union seems high. One reason for this may be the lack of a Big Bird-type satellite capable of staying in orbit for longer periods of time and performing a variety of missions. The first Big Bird satellite was launched in 1971 by the United States and now these satellites are performing most of the reconnaissance missions. The United States launched four photographic reconnaissance satellites during 1975. Of these, two were the large Big Bird satellites and the remaining two were close-look and ocean-surveillance satellites. Fewer US photographic reconnaissance satellites have been launched in recent years because the lifetimes of these satellites are increasing almost every year. The first Big Bird satellite launched in 1971, for example, had a lifetime of only 52 days compared with the lifetime of 150 days of the satellite launched on 8 June 1975.

High resolution

An analysis of the image quality of photographs taken from space suggests that a ground resolution of 15 cm is feasible. The possibility of obtaining such high resolutions is perhaps not surprising because of the improvements in lenses which have benefited from computer designs, and in photographic films with a fine grain and high sensitivity. With such a resolution it should not be difficult to observe and identify such objects as anti-ballistic missile launchers, large radar installations and intercontinental ballistic missiles as well as ballistic missile submarines which have temporarily surfaced. The restrictions on the number of strategic bombers under the recent tentative arms accord reached by the United States and the Soviet Union could also be verified by satellites.

It would be equally easy to use these means for guarding against significant concentrations of armed forces—including tanks, heavy artillery and so on—a point which may be of relevance, for example, in some arms-control settlements in Europe. There is also some indication that satellite reconnaissance might be helpful in identifying underground nuclear explosions and efforts are being made to develop sensors, mounted on satellites, to detect field tests of certain chemical weapons.

Real-time close-look reconnaissance

Although the time interval between the location of an area of interest by an area-surveillance satellite and obtaining high-resolution photographs using close-look satellites has been reduced by the use of large Big Bird satellites, this time interval is still long since the film from a Big Bird satellite must first be recovered and processed for analysis. Development of real-time close-look reconnaissance by Big Bird satellites has already begun in the United States under a project code-named 1010. The aim is to convert the high-resolution image on a photograph into electronic signals using a new-generation scanning system and then to transmit the signals back to the Earth via a data-relay satellite. A similar project is being developed to provide the United States with real-time oceanic surveillance. The availability of such techniques will give almost instantaneous surveillance of areas of potential interest.

China joins the club

Until 1975, only the Soviet Union and the United States possessed the reconnaissance satellite capability to inspect foreign territory. On 26 July 1975, the People's Republic of China launched an earth satellite (China-3) with orbital characteristics typical of a reconnaissance satellite. China's first two satellites, launched into orbit in 1970 and 1971, did not have such orbital characteristics and were thus probably not capable of reconnaissance missions.

Since the launch of China-3, China-4 and China-5 have been orbited, the latter with orbital parameters similar to those of China-3. The secrecy about the payloads and specific functions of these satellites, together with the statements made in Hsinhua News Agency reports about the satellite programme being geared to 'preparedness against war', leads one to believe that China-3 may well be the first of a series in a Chinese military reconnaissance satellite programme.

China's development of a satellite surveillance programme is not unmotivated since it has an advanced missile programme, and if its missiles are to be used as a credible deterrent, surveillance of the missile forces of other nations becomes necessary.

At present only three powers are capable of inspecting each other, as well as other nations, from space. As long as they maintain this monopoly, the application of reconnaissance satellites for arms-control agreements may be limited to their reciprocal arrangements. To be used in multilateral arms-control treaties, under which each party must obtain assurance of compliance of the treaty obligations by all other parties, reconnaissance through satellites would also have to become a multilateral undertaking.

11. *World arms trade**

The international trade in arms has grown rapidly and consistently, both in volume and in scope, since World War II. But since the October 1973 Arab-Israeli War, the growth in the arms trade can only be described as explosive. The current annual value of the trade is probably $10–12 billion and is unlikely to decline in the near future. In 1974, for example, the total value of arms export contracts signed by the USA, the USSR and France has been estimated at nearly $25 billion.

A total of 95 countries imported major weapons (such as missiles, aircraft, ships, tanks and so on) in 1975. Of the total trade, it is that with the underdeveloped countries which has attracted the most attention both because, to a large extent, it represents an extension of the conflict between East and West and because the weapons supplied have been extensively used.

---

* From *SIPRI Yearbooks*, 1974 and 1976.

In addition to horizontal proliferation, the complexity and sophistication of the weaponry being supplied is also escalating rapidly. The spread of sophisticated weaponry has obvious effects for the minimum level of conflict, should conflict break out. The amount of destruction it is possible to inflict through the possession of sophisticated weapons is clearly evidenced by the war in Viet-Nam and the recent Arab-Israeli War.

In just three years, 1973–75, the USA secured firm orders for military equipment and services valued at $13.7 billion from the OPEC countries. As this figure suggests, the United States as well as the Soviet Union, France, the United Kingdom and many of the smaller arms-producing countries found the temptation of meeting this lavish new demand for armaments utterly irresistible. And in countries that to date have pursued restrictive policies with regard to the export of armaments—FR Germany and Japan, for example—there were powerful internal pressures to liberalize these policies.

Political and economic considerations

The United States or the Soviet Union exports weapons primarily for political or military reasons, economic considerations being of secondary importance. For France and the United Kingdom the motivations are reversed and both countries vigorously promote the sale of weapons on a global scale. This is particularly true of France. Following the restrictions on oil supplies and the huge increases in the prices of oil, France was one of the first countries to attempt to secure barter arrangements with the major oil-producing countries in the Middle East, offering French technology and armaments in exchange for long-term supplies of oil.

The actual value, in current price terms, of the global traffic in weapons, equipment and related services can only be guessed at. The USA is still the only country to provide detailed information—or indeed any information at all—on its activities in this field. And although that country is currently the largest arms exporter, the combined activities of the Soviet Union, France, the United Kingdom and a host of lesser arms suppliers are certainly of the same order of magnitude. Nevertheless, the US data illustrates a number of trends that have a wider relevance, namely, the dominance of sales over grants, the comparatively rapid growth of demand in the third world and the expansion of commercial sales.

Perhaps of greater significance than the rapid escalation in the total value of the trade is the marked change that has occurred in the nature of the equipment demanded and, more generally, in the comprehensiveness of the military capability which a number of third world countries are seeking to acquire. Some third world countries are now insisting on and receiving the very latest technologies across almost the entire spectrum of conventional weapon systems and related equipment. Supplying countries have, to a significant extent, dropped their inhibitions regarding the export of highly sophisticated and/or newly developed conventional weapons and weapon subsystems.

Similarly, many of the arms contracts signed in recent years go far beyond the mere transfer of weapons to include training, technical support, the establishment of maintenance and repair facilities in the purchasing country, and construction projects. The escalation in the complexity of the systems being purchased has led to a large derived demand for technical support services to permit the fastest possible assimilation of the new equipment.

It is true that these developments are at present heavily concentrated in the major oil-producing countries, particularly Iran and Saudi Arabia. In view of the costs involved, this is hardly surprising. The more sophisticated the weapon system, the larger and more complex become the infrastructure and logistical facilities required to support it. Nevertheless the fact that weapons and equipment incorporating the newest technologies will be exported if a cash demand exists will almost certainly mean the proliferation of these items as rapidly as financial considerations permit.

Trade with the Third World

In two years—1974 and 1975—the value of resources transferred to the third world in the form of major weapons has increased by more than 60 per cent. The cumulative value of major weapon transfers in the six years 1970–75 ($19.2 billion) is already larger than that for the decade 1960–69 ($14.2 billion) and nearly three times that for the decade 1950–59 ($6.8 billion).

The Middle East is, of course, primarily responsible for the acceleration in the value of imports of major weapons by the third world but significant increases have also occurred in Africa, in the Far East (excluding Viet-Nam) and in Latin America. The fact that the transactions are predominantly on a cash or credit basis has made it easier for the supplying countries to absolve themselves of responsibly for whatever impact—military, economic or political—the arms supplied may have. Purchasing countries, it is argued, are sovereign states that are free to dispose of their resources as they see fit. This rationalization has been severely damaged by the recent disclosures of the demand-creating activities of many of the major arms manufacturers, including the payment of large commissions and in some cases outright bribes to secure contracts. Although these activities primarily affect the distribution of the demand, the major powers, because of the example they have set, must still accept a large part of the responsibility for the demand itself.

The international transfer of weapons, weapon technology and industrial know-how for weapon production has passed well beyond the point at which it could be regarded as essentially a sideline of the main arms race between East and West. In size, geographic scope and particularly in comprehensiveness, the arms trade has become a phenomenon of major importance, the control and limitation of which can only come about through a general commitment to diminish the role of relative military strength in international relations and through the pursuit of effective arms control and disarmament measures.

Appendix 1A. World military expenditure, 1954–1975

Table 1A1. World summary: constant price figures

US $ mn, at prices and 1970 exchange rates (final column, X, at current prices and exchange rates)

	1954	1955	1960	1965	1970	1973	1974	1975	1974X
USA	62 370	58 850	59 554	63 748	77 854	68 594	67 643	64 178	85 906
Other NATO	20 023	19 755	21 760	25 775	26 615	30 043	31 014	31 635	49 424
Total NATO	**82 393**	**78 605**	**81 314**	**89 523**	**104 469**	**98 637**	**98 657**	**95 813**	135 330
USSR	31 100	34 900	32 700	44 900	63 000	63 000	61 900	61 100	61 900
Other WTO[a]	2 150	2 600	2 958	4 598	7 498	8 713	9 273	10 213	9 273
Total WTO	**33 250**	**37 500**	**35 658**	**49 498**	**70 498**	**71 713**	**71 173**	**71 313**	71 173
Other Europe	2 055	2 040	2 295	2 938	3 362	3 693	3 722	3 985	6 147
Middle East	475	595	1 035	1 785	4 570	8 588	10 680	13 140	15 902
South Asia	870	935	1 030	2 166	2 236	2 619	2 500	2 545	3 397
Far East (excl. China)	1 765	1 770	2 800	4 231	5 870	7 080	7 055	7 000	10 010
China	[3 700]	[3 700]	[4 100]	[7 900]	[12 000]	[13 100]	[13 100]	[13 100]	[15 000]
Oceania	672	687	624	993	1 332	1 269	1 248	1 280	2 255
Africa (excl. Egypt)	130	150	305	970	1 918	2 025	2 135	2 750	3 146
Central America	185	210	340	466	618	700	690	700	904
South America	1 165	1 200	1 320	1 699	2 110	2 897	2 225	2 220	3 641
World total	**126 660**	**127 392**	**130 821**	**162 169**	**208 983**	**212 321**	**213 185**	**213 846**	266 905

[a] At current prices and Benoit-Lubell exchange rates.

Table 1A2. World military expenditure: growth rates and percentage distribution, 1955–1975

	Average annual per cent change	*Percentage distribution*				
		1955	1960	1965	1970	1975
World total[a]	**2.6**	**100.0**	**100.0**	**100.0**	**100.0**	**100.0**
NATO	1.0	61.7	62.3	55.2	49.9	44.8
USA	0.5	46.2	45.5	39.3	37.2	30.0
WTO	3.3	29.4	27.3	30.5	33.7	33.3
USSR	2.8	27.4	25.0	27.7	30.1	28.5
Other Europe	3.4	1.6	1.8	1.8	1.6	1.8
Other developed	4.1	1.2	1.1	1.3	1.4	1.5
China	6.5	2.9	3.1	4.9	5.7	6.1
Third World	10.3	3.2	4.6	6.3	7.7	12.3
Middle East	16.7	0.5	0.8	1.1	2.2	6.1

[a] Totals may not equal 100 because of rounding.

Table 1A3. NATO: constant price figures

US $ mn, at 1970 prices and 1970 exchange rates (final column, X, at current prices and exchange rates)

	1954	1955	1960	1965	1970	1973	1974	1975	1974X
North America:									
Canada	2 508	2 576	2 143	1 983	2 040	2 052	2 203	(2 161)	2 925
USA	62 370	58 850	59 554	63 748	77 854	68 594	67 643	(64 178)	85 906
Europe:									
Belgium	605	503	519	636	755	837	871	(940)	1 485
Denmark	249	244	264	363	368	381	417	(441)	736
France	4 217	3 922	5 158	5 658	5 919	6 373	6 381	(6 720)	10 080
FR Germany	2 603	2 968	4 375	6 232	6 188	7 363	7 688	(7 753)	13 853
Greece	166	170	209	237	474	533	510	(817)	804
Italy	1 438	1 428	1 678	2 254	2 506	3 126	3 128	(2 777)	4 387
Luxembourg	16	17	7	11	8	10	11	(11)	18
Netherlands	789	827	720	959	1 103	1 246	1 301	(1 335)	2 405
Norway	285	238	230	338	389	401	412	(430)	715
Portugal	125	132	163	316	436	416	499	(461)	989
Turkey	328	351	401	532	579	438	808	(1 212)	1 135
UK	6 694	6 379	5 893	6 256	5 850	6 567	6 785	(6 577)	9 892
Total NATO	**82 393**	**78 605**	**81 314**	**89 523**	**104 469**	**98 637**	**98 657**	**95 813**	135 330
Total NATO (excl USA)	**20 023**	**19 755**	**21 760**	**25 775**	**26 615**	**30 043**	**31 014**	**31 635**	49 424
Total NATO Europe	**17 515**	**17 179**	**19 617**	**23 792**	**24 595**	**27 991**	**28 811**	**29 474**	46 499

Table 1A4. NATO: military expenditure as a percentage of gross domestic product

Per cent

	1954	1955	1960	1965	1970	1971	1972	1973	1974
North America:									
Canada	7.0	6.6	4.3	3.0	2.4	2.3	2.1	2.0	2.0
USA	11.6	10.0	8.9	7.5	7.9	7.1	6.7	6.0	6.1
Europe:									
Belgium	4.8	3.8	3.4	3.2	2.9	2.8	2.8	2.7	2.8
Denmark	3.2	3.2	2.7	2.8	2.4	2.5	2.3	2.1	2.4
France	7.3	6.4	6.4	5.2	4.2	4.0	3.9	3.8	(3.8)
FR Germany	4.0	4.1	4.0	4.3	3.3	3.3	3.4	3.4	3.6
Greece	5.5	5.2	4.9	3.6	4.8	4.7	4.6	4.2	4.2
Italy	4.0	3.7	3.3	3.3	2.7	2.9	3.1	3.0	2.9
Luxembourg	3.3	3.2	1.1	1.4	0.8	0.8	0.9	0.8	–
Netherlands	6.0	5.7	4.1	3.9	3.5	3.4	3.4	3.4	3.5
Norway	5.0	3.9	3.2	3.7	3.5	3.4	3.3	3.1	3.1
Portugal	4.2	4.2	4.2	6.2	7.0	7.4	6.9	5.9	–
Turkey	5.4	5.1	4.7	4.8	4.3	4.5	4.3	4.2	3.9
UK	8.8	8.2	6.5	5.9	4.9	5.0	5.2	5.0	–

Table 1A5. WTO: current price figures

US $ mn, at Benoit–Lubell exchange rates

	1954	1955	1960	1965	1970	1972	1973	1974	1975
Bulgaria	–	–	154	198	279	337	364	416	472
Czechoslovakia	918	1 227	1 033	1 191	1 755	1 976	1 976	2 035	2 271
German DR	–	–	295	914	2 006	2 242	2 457	2 625	2 821
Hungary	–	–	179	332	567	543	547	611	649
Poland	666	791	937	1 461	2 142	2 324	2 538	2 676	2 971
Romania	–	–	360	502	749	818	831	910	1 029
USSR[a]	31 100	34 900	32 700	44 900	63 000	63 000	63 000	61 900	61 100
Total WTO	**[33 250]**	**[37 500]**	**35 658**	**49 498**	**70 498**	**71 240**	**71 713**	**71 173**	**71 313**

[a] At SIPRI-estimated exchange rates (see *SIPRI Yearbook*, 1974, pp. 191 ff.).

Table 1A6. WTO: military expenditure as a percentage of net material product

Per cent

	1954	1955	1960	1965	1970	1971	1972	1973	1974
Bulgaria	–	–	4.0	3.5	3.1	3.4	3.5	3.5	3.7
Czechoslovakia	6.3	7.8	5.4	5.9	4.8	5.0	4.9	4.7	4.5
German DR	–	–	1.4	3.7	(6.2)	(6.3)	(6.3)	[6.5]	–
Hungary	–	–	2.2	3.4	3.6	3.4	3.0	2.7	2.9
Poland	4.2	5.6	4.0	4.4	4.6	4.3	3.9	3.8	3.5
USSR	10.9	11.4	6.4	6.6	6.2	5.9	5.7	5.3	5.0

Table 1A7. Other Europe: constant price figures

US $ mn, at 1970 prices and 1970 exchange rates (final column, X, at current prices and exchange rates)

	1954	1955	1960	1965	1970	1973	1974	1975	1974X
Albania[a]	–	–	–	73	120	148	154	160	154
Austria	3	12	104	135	160	166	170	(193)	309
Finland	64	86	103	134	142	179	162	(190)	271
Ireland	38	35	35	43	51	70	72	–	110
Spain	324	310	349	431	603	741	746	–	1 362
Sweden	758	781	833	1 118	1 190	1 210	1 162	(1 201)	1 818
Switzerland	237	255	297	435	467	479	455	(463)	903
Yugoslavia	584	512	514	569	629	700	801	(955)	1 220
Total Other Europe	**[2 055]**	**[2 040]**	**[2 295]**	**2 938**	**3 362**	**3 693**	**3 722**	**[3 985]**	6 147

[a] Figures for Albania are at current prices and Benoit–Lubell exchange rates.

Table 1A8. Other Europe: military expenditure as a percentage of gross domestic product

Per cent

	1954	1955	1960	1965	1970	1971	1972	1973	1974
Austria	0.1	0.2	1.2	1.2	1.1	1.0	1.0	1.0	0.9
Finland	1.4	1.6	1.7	1.7	1.4	1.5	1.5	1.4	1.2
Ireland	1.7	1.6	1.4	1.4	1.3	1.4	1.5	1.5	1.6
Spain	2.4	2.2	2.2	1.8	1.9	1.8	1.8	1.9	–
Sweden	4.9	4.8	4.0	4.1	3.6	3.7	3.7	3.4	3.2
Switzerland	2.7	2.8	2.5	2.7	2.3	2.3	2.1	2.0	2.0
Yugoslavia[a]	12.6	10.3	7.2	5.4	5.0	4.4	4.8	4.6	–

[a] Percentage of gross material product.

Table 1A9. Middle East: constant price figures

US $ mn, at 1970 prices and 1970 exchange rates (final column, X, at current prices and exchange rates)

	1954	1955	1960	1965	1970	1972	1973	1974	1975	1974X
Cyprus	–	–	–	9	8	(7)	(7)	–	–	(10)[a]
Egypt	166	251	[264]	501	1 263	1 420	2 327	2 315	(2 114)	3 136
Iran	78	107	216	323	714	990	1 360	3 050	4 012	4 748
Iraq	75	67	147	268	401	394	547	540	–	806
Israel	32	34	144	288	1 278	1 375	2 415	1 972	(1 950)	3 250
Jordan	(40)	(41)	(68)	(71)	105	109	95	83	(77)	138
Kuwait[b]	–	–	–	31	67	88	314	[518]	[588]	[632]
Lebanon	10	12	17	31	43	61	67	85	(74)	149
Oman[b]	–	–	–	–	–	[48]	91	243	300	292
Saudi Arabia	–	–	–	(138)	446	(680)	(1 010)	(1 420)	(2 943)	2 103
Syria	28	30	78	113	162	180	289	358	385	538
United Arab Emirates[b]	–	–	–	–	–	–	13	19	60	23
Yemen[b]	–	–	–	[2]	13	22	29	40	–	48
Yemen, Democratic[b]	–	–	–	–	19	23	24	–	–	29[a]
Total Middle East	[475]	[595]	[1 035]	[1 785]	[4 570]	5 407	8 588	[10 680]	[13 140]	15 902

[a] 1973. [b] At current prices and 1970 exchange rates.

Table 1A10. Middle East: military expenditure as a percentage of gross domestic product

Per cent

	1954	1955	1960	1965	1970	1971	1972	1973	1974
Cyprus	–	–	–	2.4	1.3	1.4	1.2	–	–
Egypt	–	–	5.6	7.7	18.0	20.1	19.2	31.4	–
Iran	–	–	4.2	4.7	6.3	5.4	6.6	6.7	–
Iraq	4.7	4.1	7.1	9.2	11.1	10.2	–	–	–
Israel	2.8	2.5	6.6	7.9	23.6	22.8	20.5	33.3	–
Jordan	–	–	19.4	12.8	17.8	18.2	17.7	16.1	14.7
Kuwait	–	–	–	1.5	2.5	2.2	2.1	5.7	[6.3]
Lebanon	–	–	–	2.6	2.8	2.6	3.3	–	–
Saudi Arabia	–	–	–	5.7	10.0	8.9	9.4	7.1	–
Syria	–	–	–	7.9	9.6	8.4	8.2	14.9	13.8
Yemen	–	–	–	–	3.1	3.2	3.6	4.3	–
Yemen, Democratic	–	–	–	–	13.7	–	–	–	–

Table 1A11. South Asia: constant price figures

US $ mn, at 1970 prices and 1970 exchange rates (final column, X, at current prices and exchange rates)

	1954	1955	1960	1965	1970	1973	1974	1975	1974X
Afghanistan	–	–	–	44.4	30.2	42.3	–	–	42[a]
Bangla Desh	–	–	–	–	–	40.2	40.9	(35.6)	81
India	585.6	610.2	677.6	1 567.6	1 558.2	1 786.0	1 757.0	(1 793.0)	2 726
Nepal	–	–	[3.1]	[3.8]	5.8	6.9	6.9	(7.8)	9
Pakistan	240.9	281.3	290.1	537.7	623.0	723.0	629.0	(635.0)	520
Sri Lanka	7.2	6.6	16.0	12.8	19.0	20.3	16.1	(19.9)	19
Total South Asia	**[870.0]**	**[935.0]**	**[1 030.0]**	**2 166.3**	**2 236.2**	**2 618.7**	**[2 500.0]**	**[2 545.0]**	3 397

[a] 1973.

Table 1A12. South Asia: military expenditure as a percentage of gross domestic product

Per cent

	1954	1955	1960	1965	1970	1971	1972	1973	1974
India	[1.8]	[1.7]	[1.9]	3.6	3.0	3.4	3.5	3.4	–
Nepal	–	–	–	[0.4]	0.7	0.7	0.7	0.7	–
Pakistan	[3.1]	[3.4]	2.8	4.0	[3.7]	[4.4]	7.2	6.0	–
Sri Lanka	0.6	0.5	1.1	0.8	0.9	1.3	1.1	0.8	0.6

Table 1A13. Far East: constant price figures

US $ mn, at 1970 prices and 1970 exchange rates (final column, X, at current prices and exchange rates)

	1954	1955	1960	1965	1970	1973	1974	1975	1974X
Brunei[a]	–	–	–	9.5	16.5	12.0	19.4	43.4	25
Burma[a]	77.4	70.8	89.3	107.0	121.9	117.1	(128.8)	(156.0)	127
Cambodia	–	–	41.5	41.4	124.8	145.5	–	–	110[b]
Indonesia	224.0	182.0	336.0	127.0	301.0	361.0	[415.0]	(512.0)	775
Japan	843.6	795.3	798.3	1 095.6	1 594.8	2 128.0	1 946.0	(2 056.0)	3 670
Korea, North	–	–	–	[350.0]	(745)	1 068.0	1 307.0	–	1 574
Korea, South	141.3	113.6	172.5	170.6	324.6	443.1	420.5	(551.8)	518
Laos	–	–	–	41.2	38.0	32.0	25.3	–	25
Malaysia	64.4	57.8	46.6	105.0	165.0	190.0	178.0	(231.0)	311
Mongolia[a]	–	–	–	[15.0]	[38.0]	53.0	90.0	93.0	108
Philippines	47.4	46.4	51.5	49.0	85.0	123.0	114.0	(165.0)	211
Singapore	–	–	–	–	100.6	127.3	126.1	(151.0)	251
Taiwan	–	153.2	226.0	370.0	482.0	(638.0)	[560.0]	–	(903)
Thailand	65.5	56.3	80.2	103.7	210.0	248.0	234.0	[262.0]	351
Viet-Nam, North	–	–	–	[620.0]	[585.0]	[520.0]	–	–	[520][c]
Viet-Nam, South	–	–	386.0	1 026.0	938.0	874.0	740.0	–	531
Total Far East	**[1 765.0]**	**[1 770.0]**	**[2 800.0]**	**4 231.0**	**5 870.2**	**[7 080.0]**	**[7 055.0]**	**[7 000.0]**	10 010

[a] At current prices and 1970 exchange rates. [b] 1972. [c] 1973.

Table 1A14. Far East: military expenditure as a percentage of gross domestic product

Per cent

	1954	1955	1960	1965	1970	1971	1972	1973	1974
Burma	6.7	5.9	6.0	(6.6)	5.7	5.7	5.3	4.5	–
Cambodia	–	–	–	6.1	–	–	–	–	–
Indonesia	–	–	5.4	1.3	3.1	3.2	3.2	2.7	–
Japan	2.1	1.8	1.1	0.9	0.8	0.8	0.9	0.8	–
Korea, South	6.6	5.1	6.0	3.7	3.9	4.3	4.4	3.7	3.1
Malaysia	–	3.2	2.2	4.0	5.1	5.5	5.2	4.7	4.4
Philippines	1.8	1.7	1.4	1.1	1.2	1.1	1.3	1.6	2.3
Singapore	–	–	–	–	5.4	5.9	5.3	4.8	4.7
Taiwan	–	9.3	12.9	[11.3]	8.8	9.6	9.3	8.4	–
Thailand	2.8	2.4	2.6	2.3	3.3	3.7	3.5	2.9	2.7
Viet-Nam, South	–	–	6.6	21.2	16.5	16.2	20.9	16.4	–

Table 1A15. Oceania: constant price figures

US $ mn, at 1970 prices and 1970 exchange rates (final column, X, at current prices and exchange rates)

	1954	1955	1960	1965	1970	1973	1974	1975	1974X
Australia	577.0	598.0	534.0	882.0	1 200.0	1 147.0	1 127.0	1 160.0	2 041
New Zealand	94.7	89.2	89.8	111.0	132.0	122.0	121.0	120.0	214
Total Oceania	**671.7**	**687.2**	**623.8**	**993.0**	**1 332.0**	**1 269.0**	**1 248.0**	**1 280.0**	**2 255**

Table 1A16. Oceania: military expenditure as a percentage of gross domestic product

Per cent

	1954	1955	1960	1965	1970	1971	1972	1973	1974
Australia	3.6	3.6	2.6	3.4	3.4	3.3	3.1	2.7	2.6
New Zealand	2.7	2.5	2.1	2.1	2.2	2.0	1.8	1.7	–

Table 1A17. Africa: constant price figures

US $ mn, at 1970 prices and 1970 exchange rates (final column, X, at current prices and exchange rates)

	1954	1955	1960	1965	1970	1973	1974	1975	1974X
Algeria[a]	–	–	–	99	99	110	177	(211)	210
Benin (Dahomey)[a]	–	–	–	(3.6)	4.3	5.1	–	–	6[c]
Burundi	–	–	–	2.4	3.1	–	–	–	4[b]
Cameroon	–	–	11.3	15.3	19.8	(23.8)	–	–	38[c]
Central African Republic	–	–	–	2.3	4.9	4.8	4.5	–	7
Chad	–	–	–	3.7	12.6	–	–	–	[15][b]
Congo	–	–	–	5.3	[10.1]	[11.3]	13.3	–	19
Ethiopia	–	–	24.9	53.8	34.4	37.4	35.5	(46.6)	48
Gabon	–	–	–	3.1	4.6	6.7	7.2	–	11
Ghana	–	8.9	30.6	29.7	42.0	37.5	44.5	–	64
Guinea[a]	–	–	–	11.1	[18.0]	–	–	–	19[b]
Ivory Coast	–	–	–	14.5	17.6	19.7	–	–	27[c]
Kenya	–	–	3.1	10.8	17.1	31.9	(30.7)	(27.0)	42
Liberia	–	–	–	3.5	3.8	3.0	–	–	4[c]
Libya	–	–	(5.9)	25.9	[365.0]	[400.0]	[290.0]	–	400
Malagasy Rep.	–	–	1.9	10.7	12.1	[13.1]	12.5	–	21
Malawi	–	–	–	(1.1)	1.4	2.8	–	–	3[c]
Mali[a]	–	–	–	4.3	6.1	8.4	–	–	11[c]
Mauritania	–	–	–	2.3	2.4	–	–	–	3[d]
Mauritius	–	–	0.3	0.3	0.4	0.5	–	–	1[c]
Morocco	–	–	52.1	65.1	87.7	116.8	110.4	–	187
Niger	–	–	–	2.3	3.8	[3.6]	(3.2)	–	5
Nigeria	[7.6]	7.3	26.3	51.9	434.0	289.0	307.0	(723.0)	492
Rhodesia, S.	–	–	–	19.3	25.5	36.5	51.4	(56.6)	72
Rwanda	–	–	–	2.6	4.5	[4.4]	–	–	6[c]
Senegal	–	–	–	14.6	16.1	14.1	13.9	(14.2)	22
Sierra Leone	–	–	–	2.7	3.7	3.9	–	–	4[c]
Somalia	–	–	–	5.5	11.2	[13.1]	11.5	–	16
South Africa	85	86	81	300	360	497	654	(792)	950
Sudan	10.3	11.7	24.6	49.9	93.3	84.7	64.9	(50.5)	109
Tanzania	–	–	–	8.3	24.5	32.7	28.4	–	46
Togo	–	–	–	(2.9)	3.0	3.8	4.3	–	7
Tunisia	–	–	18.6	16.3	22.5	25.3	28.3	(36.4)	40
Uganda	–	3.0	1.6	13.1	26.6	36.1	21.0	–	49
Upper Volta	–	–	1.5	3.4	4.2	[4.0]	–	–	6[c]
Zaire	–	–	–	86.9	96.0	[62.0]	48.9	–	104
Zambia	–	–	10.4	23.0	22.5	52.0	54.0	–	78
Total Africa	**[130.0]**	**[150.0]**	**[305.0]**	**970.2**	**1 917.8**	**[2 025]**	**[2 135]**	**[2 750]**	3 146

[a] At current prices and 1970 exchange rates. [b] 1972. [c] 1973. [d] 1971.

Table 1A18. Africa: military expenditure as a percentage of gross domestic product

Per cent

	1954	1955	1960	1965	1970	1971	1972	1973	1974
Algeria	–	–	–	3.5	2.1	2.1	1.8	1.8	–
Benin (Dahomey)	–	–	–	2.2	[2.1]	2.1	[2.0]	2.0	–
Burundi	–	–	–	–	(1.5)	–	–	–	–
Cameroon	–	–	–	2.1	1.9	–	–	–	–
Central African Republic	–	–	–	(1.3)	2.4	2.6	–	–	–
Chad	–	–	–	–	(4.7)	–	–	–	–
Congo	–	–	–	[2.9]	[4.4]	[4.0]	–	–	–
Ethiopia	–	–	1.7	3.2	1.9	1.9	1.9	1.8	–
Gabon	–	–	–	1.5	1.4	[1.4]	1.6	–	–
Ghana	–	0.6	1.6	1.6	1.9	1.7	1.6	–	–
Guinea	–	–	–	–	–	–	–	–	–
Ivory Coast	–	–	–	1.3	1.2	1.2	1.1	1.1	–
Kenya	–	–	0.4	1.0	1.1	1.2	1.5	1.7	(1.6)
Liberia	–	–	–	0.9	0.9	1.0	0.8	–	–
Libya	–	–	–	1.4	[9.8]	[8.3]	[7.8]	[6.8]	–
Malagasy Rep.	–	–	0.3	1.6	1.4	1.4	1.3	–	–
Malawi	–	–	–	0.4	0.4	0.4	0.4	0.6	–
Mauritania	–	–	–	(1.4)	[1.3]	[1.3]	–	–	–
Mauritius	–	–	0.2	0.2	0.2	0.2	0.2	0.2	–
Morocco	–	–	2.3	2.4	2.6	2.7	2.8	3.1	3.1
Niger	–	–	–	0.7	0.9	–	–	–	–
Nigeria	[0.2]	0.2	0.5	0.8	5.9	4.3	4.2	2.9	–
Rhodesia	–	–	–	1.7	1.7	1.6	1.5	1.9	2.4
Rwanda	–	–	–	–	2.2	2.3	2.5	2.3	–
Senegal	–	–	–	2.0	1.9	2.1	1.9	–	–
Sierra Leone	–	–	–	0.7	0.9	0.9	–	–	–
Somalia	–	–	–	–	–	–	–	–	–
South Africa	1.0	1.0	0.8	2.3	2.1	2.4	2.1	2.3	2.9
Sudan	–	–	1.6	3.0	5.2	5.5	–	–	–
Tanzania	–	–	–	0.8	1.9	2.4	2.2	2.4	–
Togo	–	–	–	1.6	1.1	1.2	1.2	–	–
Tunisia	–	–	2.2	1.4	1.6	1.4	1.4	1.3	1.1
Uganda	–	0.5	0.3	1.3	2.0	3.6	–	–	–
Upper Volta	–	–	(0.7)	1.5	–	–	–	–	–
Zaire	–	–	–	–	5.0	4.6	4.4	–	–
Zambia	–	–	1.1	1.8	1.3	4.2	5.4	–	–

[a] GDP figure used excludes Eastern states. [b] GDP at factor cost.

Table 1A19. Central America: constant price figures

US $ mn, at 1970 prices and 1970 exchange rates (final column, X, at current prices and exchange rates)

	1954	1955	1960	1965	1970	1972	1973	1974	1974X
Cuba[a]	–	–	–	215	290	[320]	–	–	[347][b]
Dominican Rep.	–	–	39.9	36.9	31.3	31.0	28.6	24.9	36
El Salvador	6.5	7.2	6.5	10.0	10.6	14.1	13.6	–	15[c]
Guatemala	7.4	8.6	10.2	15.4	28.7	19.5	18.3	18.1	24
Haiti	6.5	6.4	8.7	8.0	7.2	6.9	5.7	5.3	8
Honduras	4.2	3.9	5.2	6.6	8.6	14.3	14.0	13.1	17
Jamaica	–	–	–	5.3	5.5	6.7	–	–	8[b]
Mexico	64.2	73.4	106.6	157.2	220.2	270.0	269.2	252.5	424
Nicaragua	–	–	–	9.1	12.1	(14.8)	(15.0)	[16.4]	21
Trinidad & Tobago	–	–	–	2.6	3.8	3.5	3.2	–	4
Total Central America	**[185.0]**	**[210.0]**	**[340.0]**	**466.1**	**618.0**	**700.8**	**[700.0]**	**[690.0]**	904

[a] At current prices and 1970 exchange rates. [b] 1972. [c] 1973.

Table 1A20. Central America: military expenditure as a percentage of gross domestic product

Per cent

	1954	1955	1960	1965	1970	1971	1972	1973	1974
Cuba	–	–	–	–	–	–	–	–	–
Dominican Republic	–	–	4.6	3.7	2.1	1.9	1.7	1.6	–
El Salvador	–	–	1.1	1.2	1.0	1.1	1.2	1.1	-
Guatemala	0.9	1.0	0.9	1.1	1.5	0.9	0.9	0.8	0.8
Haiti	–	–	[2.2]	–	1.4	1.4	1.4	–	–
Honduras	1.1	1.0	1.1	1.2	1.2	1.5	1.9	1.8	1.7
Jamaica	–	–	–	0.5	0.4	0.5	0.5	–	–
Mexico	0.6	0.6	0.7	0.7	0.7	0.7	0.7	0.7	0.7
Nicaragua	–	–	–	1.3	1.5	1.4	1.7	1.7	–
Trinidad & Tobago	–	–	–	0.3	0.4	–	–	–	–

Table 1A21. South America: constant price figures

US $ mn, at 1970 prices and 1970 exchange rates (final column, X, at current prices and exchange rates)

	1954	1955	1960	1965	1970	1973	1974	1975	1974X
Argentina	428.9	341.3	406.3	391.7	449.8	349.0	376.0	(226.0)	1 278
Bolivia	–	5.6	[7.1]	20.0	19.2	33.6	24.6	–	35
Brazil	394	450	462	697	853	1 428	865	(977)	1 199
Chile	79.7	119.2	98.4	98.0	207.0	402.3	235.0	–	182
Colombia	67.0	66.3	49.5	106.0	101.6	87.4	67.8	(72.4)	91
Ecuador	19.0	22.0	25.4	26.6	38.0	47.8	(27.0)	–	35
Guyana	–	–	–	–	3.8	4.1	–	–	5[a]
Paraguay	–	–	–	[8.2]	12.0	14.3	13.0	(15.9)	21
Peru	55.8	59.6	86.3	135.4	179.8	210.1	175.0	–	257
Uruguay	–	–	–	37.5	47.7	56.9	–	–	70[a]
Venezuela	73.9	92.5	139.7	178.5	198.0	263.0	372.0	352.0	468
Total South America	[1 165.0]	[1 200.0]	[1 320.0]	**1 698.9**	**2 109.9**	**2 896.5**	[2 225.0]	[2 220.0]	3 641

[a] 1973.

Table 1A22. South America: military expenditure as a percentage of gross domestic product

Per cent

	1954	1955	1960	1965	1970	1971	1972	1973	1974
Argentina	2.9	2.2	2.3	1.8	1.9	1.6	1.6	1.3	–
Bolivia	–	0.3	[1.1]	2.5	1.9	1.8	2.3	2.7	1.9
Brazil	2.2	2.3	2.0	2.5	1.9	2.4	2.2	2.3	–
Chile	2.2	3.3	2.6	2.0	2.5	2.3	2.6	3.6	–
Colombia	2.2	2.1	1.2	2.0	1.4	2.5	1.2	1.0	–
Ecuador	2.4	2.7	2.4	2.1	2.2	1.9	2.0	2.0	1.0
Guyana	–	–	–	–	1.4	1.2	1.2	1.5	–
Paraguay	–	–	–	[1.7]	2.0	1.3	2.2	1.9	1.6
Peru	2.1	2.1	2.4	2.9	2.9	3.4	3.3	2.9	–
Uruguay	–	–	–	1.7	1.9	2.6	2.3	2.4	–
Venezuela	1.6	1.9	2.1	2.0	1.7	2.0	2.0	1.7	1.6

Appendix 1B

Table 1B1. US and Soviet strategic nuclear forces, 1967–1976

Mid-year (1 July) figures

		Introduced	*Range, nm*	*Payload*	1967	1968	1969	1970	1971	1972	1973	1974	1975	1976
Delivery vehicles														
Strategic bombers														
USA	B-52C/D/E/F	1956	10 000	27 210 kg	(334)	(283)	(218)	(206)	(206)	(167)	(150)	(150)	(150)	80
	B-52G/H	1959	10 860	34 015 kg	283	283	283	283	283	282	274	274	274	274
	B-58	1960	(2 000)	5 442 kg	80	80	80	–	–	–	–	–	–	–
	FB-111	1970	3 300	16 780 kg	–	–	–	(28)	(76)	76	76	76	76	76
USSR	Mya-4 'Bison'	1955	5 255	9 070 kg	55	50	40	40	40	40	40	40	40	40
	Tu-20 'Bear'	1956	6 775	18 140 kg	100	100	100	100	100	100	100	100	100	100
	Tu- – 'Backfire'	1975	(3 000)	(20 000 kg)	–	–	–	–	–	–	–	–	(20)	(60)
			Bomber total:	**USA**	**697**	**646**	**581**	**517**	**565**	**525**	**500**	**500**	**500**	**430**
				USSR	**155**	**150**	**140**	**140**	**140**	**140**	**140**	**140**	**160**	**180**
Strategic submarines														
USA	With Polaris A-2	1962	n.a.	16 × A-2	13	13	13	8	8	8	8	6	3	–
	With Polaris A-3	1964	n.a.	16 × A-3	28	28	28	32	26	21	13	13	13	13
	With Poseidon C-3	1970	n.a.	16 × C-3	–	–	–	1	7	12	20	22	25	28
USSR	'Hotel' class	1960	n.a.	3 × 'SS-N-5'	9	9	9	8	8	8	8	8	8	8
	'Yankee' class	1968	n.a.	16 × 'SS-N-6'	–	(2)	(8)	(14)	(21)	(27)	(33)	34	34	34
	'Delta I' class	1973	n.a.	12 × 'SS-N-8'	–	–	–	–	–	–	(1)	(8)	(11)	(11)
	'Delta II' class	1976	n.a.	16 × 'SS-N-8'	–	–	–	–	–	–	–	–	–	(1)
			Submarine total:	**USA**	**41**	**41**	**41**	**41**	**41**	**41**	**41**	**41**	**41**	**41**
				USSR	**9**	**11**	**17**	**22**	**29**	**35**	**42**	**50**	**53**	**54**
SLBMs (*Submarine-launched ballistic missiles*)														
USA	Polaris A-2	1962	1 520	1 × 1 mt	208	208	208	128	128	128	128	96	48	–
	Polaris A-3	1964	2 500	3 × 200 kt (MRV)	448	448	448	512	416	336	208	208	208	208
	Poseidon C-3	1970	2 500	14 × 40 kt (MIRV)	–	–	–	16	112	192	320	352	400	448
USSR	'SS-N-5'	1963	700	1 × 1 mt	27	27	27	24	24	24	24	24	24	24
	'SS-N-6 mod. 1' 'SS-N-6 mod. 2'	1968 1974	1 300 1 600	1 × 1 mt 1 × 1 mt	–	32	128	224	336	432	528	544	544	544
	'SS-N-8'	1973	4 200	1 × 1 mt	–	–	–	–	–	–	12	96	132	148
			SLBM total:	**USA**	**656**	**656**	**656**	**656**	**656**	**656**	**656**	**656**	**656**	**656**
				USSR	**27**	**59**	**155**	**248**	**360**	**456**	**564**	**664**	**700**	**716**

ICBMs (*Intercontinental ballistic missiles*)														
USA	Titan II	1962	6 300	1×10 mt	54	54	54	54	54	54	54	54	54	54
	Minuteman I	1962	6 515	1×1 mt	700	600	500	490	390	290	(190)	(100)	–	–
	Minuteman II	1966	6 950	1×2 mt	300	400	500	500	500	500	(500)	(500)	450	450
	Minuteman III	1970	7 020	3×200 kt (MIRV)	–	–	–	10	110	210	(310)	(400)	550	550
USSR	'SS-7 Saddler'	1962	6 000	1×5 mt	200	200	200	200	190	190	190	190	190	90
	'SS-8 Sasin'	1963	6 000	1×5 mt	20	20	20	20	19	19	19	19	19	19
	'SS-9 Scarp'	1965	6 515	1×20 mt	(160)	(190)	(230)	288	288	288	288	288	288	213
	'SS-11 mod. 1'	1966	5 650	1×1 mt	(340)	(470)	(720)	(950)	970	970	970	970	970	870
	'SS-13 Savage'	1968	4 350	1×1 mt	–	(20)	(30)	(40)	60	60	60	60	60	60
	'SS-11 mod. 3'	1973	5 650	3×200 kt (MRV)	–	–	–	–	–	–	20	40	60	60
	'SS-18 mod. 1'	1976	5 500	1×20 mt	–	–	–	–	–	–	–	–	–	(75)
	'SS-19'	1976	5 500	6×1 mt (MIRV)	–	–	–	–	–	–	–	–	–	(100)
ICBM total:				**USA**	**1 054**	**1 054**	**1 054**	**1 054**	**1 054**	**1 054**	**1 054**	**1 054**	**1 054**	**1 054**
				USSR	**720**	**900**	**1 200**	**1 498**	**1 527**	**1 527**	**1 547**	**1 567**	**1 587**	**1 507**
Total, bombers and missiles:				**USA**	**2 407**	**2 356**	**2 291**	**2 227**	**2 275**	**2 235**	**2 210**	**2 210**	**2 210**	**2 210**
				USSR	**902**	**1 109**	**1 495**	**1 886**	**2 027**	**2 123**	**2 251**	**2 371**	**2 447**	**2 403**
Nuclear warheads														
Independently targetable warheads on missiles, SIPRI estimates														
				USA	1 710	1 710	1 710	1 938	3 386	4 626	6 490	7 086	8 010	8 634
				USSR	747	959	1 355	1 746	1 887	1 983	2 111	2 231	2 287	3 353
Total warheads on bombers and missiles, official US estimates														
				USA	4 500	4 200	4 200	4 000	4 600	5 700	6 784	7 650	8 500	8 900
				USSR	1 000	1 100	1 350	1 800	2 100	2 500	2 200	2 500	2 500	3 500

For sources and notes, see *SIPRI Yearbook*, 1976.

D

Appendix 1C. Values of arms trade with Third World countries, 1950–1975

Table 1C1. Values of imports of major weapons by third world countries: by region, 1950–1975[a]

US $ mn, at constant (1973) prices. A = yearly figures, B = five-year moving averages

Region		1950	1955	1960	1965	1970	1973	1974	1975	**Total[b]**
Far East (excl. Viet-Nam)	A	147	222	583	260	207	231	190	489	**6 850**
	B[c]	–	209	382	266	260	270	–	–	
South Asia	A	44	108	205	163	229	221	285	136	**4 953**
	B	–	147	241	180	278	267	–	–	
Middle East	A	35	186	123	337	1 118	1 704	2 260	2 696	**16 219**
	B	–	197	240	417	1 035	1 753	–	–	
North Africa	A	–	–	9	62	92	111	174	582	**1 698**
	B	–	–	12	63	89	218	–	–	
Sub-Saharan Africa	A	–	12	27	72	95	142	299	177	**1 481**
	B	–	10	31	59	72	148	–	–	
South Africa	A	8	15	4	142	59	28	210	137	**1 213**
	B	–	25	11	86	40	90	–	–	
Central America	A	6	18	45	14	4	43	90	105	**1 035**
	B	–	12	92	29	16	55	–	–	
South America	A	54	195	139	84	113	367	406	482	**3 777**
	B	–	128	111	76	160	300	–	–	
Total (excl. Viet-Nam)	A	**294**	**755**	**1 135**	**1 135**	**1 916**	**2 711**	**3 769**	**4 788**	37 281
	B	–	**730**	**1 121**	**1 175**	**1 923**	**3 074**	–	–	
Viet-Nam	A	–	9	24	57	331	63	142	15	**3 350**
	B	–	–	39	146	434	294	–	–	
Total[b]	A	**294**	**765**	**1 159**	**1 192**	**2 247**	**2 909**	**4 070**	**4 843**	40 632
	B	–	**737**	**1 160**	**1 320**	**2 385**	**3 395**	–	–	–

[a] The values include licensed production. [b] Items may not add up to totals because of rounding. Figures are rounded to nearest 10. [c] Five-year moving averages are calculated from the year arms imports began, as a more stable measure of the trend in arms imports than the often erratic year-to-year figures.

Source: SIPRI worksheets. Information on individual countries and arms transactions is available on request.

Table 1C2. Values of exports of major weapons to regions listed in table 1C1: by supplier, 1950–75[a]

US $ mn, at constant (1973) prices

Country	1950	1955	1960	1965	1970	1972	1973	1974	1975
USA	91	305	545	413	962	958	885	1 200	1 769
USSR	25	66	165	408	836	726	1 542	1 540	1 652
UK	96	175	196	203	142	283	242	481	503
France	3	70	37	74	156	269	411	357	477
Canada	14	1	11	14	28	30	3	80	5
China	23	–	125	7	17	120	21	★	48
Czechoslovakia	–	43	45	3	24	10	1	11	5
FR Germany	★	7	23	10	1	37	2	101	118
Italy	7	2	7	5	33	39	4	106	65
Japan	–	–	–	5	★	–	–	2	–
Netherlands	35	85	1	17	7	20	30	25	32
Sweden	★	6	1	–	–	4	1	5	16
Other indus. West	–	5	1	23	3	10	16	9	10
Other indus. East	–	–	★	★	–	–	13	–	2
Third world	–	1	3	3	6	14	16	211	141
Total (incl. Viet-Nam	**294**	**765**	**1 159**	**1 192**	**2 247**	**2 673**	**2 909**	**4 070**	**4 843**

[a] The values include licensed production. [b] Items may not add up to totals owing to rounding.
★ < $1 mn.

Source: SIPRI worksheets. Information on individual countries and arms transactions is available on request.

Table 1C3. Increase in the value of imports of major weapons by the third world

US $ mn, at constant (1973) prices

	Average annual imports		
	1964–68	1969–73	1974–75
Total third world	1 521	2 527	4 387
Middle East	549	1 181	2 451
Far East, total	469	640	408
excl. Viet-Nam	259	266	332
South Asia	191	277	210
North Africa	70	98	378
Sub-Saharan Africa	130	126	401
Latin America	114	206	497

Table 1C4. Distribution of the value of imports of major weapons by the third world

Per cent

Region	1970	1971	1972	1973	1974	1975
Middle East	49.7	47.4	31.0	61.0	56.4	56.1
Far East	23.9	23.0	39.2	10.6	8.5	10.5
Viet-Nam	14.7	11.7	34.5	2.3	3.6	0.3
South Asia	10.2	13.4	11.8	8.0	7.3	2.8
Latin America	5.2	7.3	9.9	4.3	10.4	12.2
North Africa	4.1	3.3	4.8	4.0	4.4	12.1
Sub-Saharan Africa	6.7	5.5	3.3	5.2	12.9	6.2
OPEC[a]	15.2	21.2	16.0	19.2	29.4	48.0

[a] Organization of Petroleum Exporting Countries. Totals will add to more than 100 because OPEC countries are situated in the Middle East, the Far East, Africa and South America.

Table 1C5. Ten leading importers of major weapons in the third world

US $ mn, at constant (1973) prices

Country	*Cumulative imports* 1965–75
1. Iran	3 220
2. Egypt	3 047
3. Syria	2 185
4. Israel	2 118
5. India	1 901
6. Viet-Nam, N.	1 513
7. Viet-Nam, S.	1 495
8. South Africa	1 061
9. Iraq	1 060
10. Libya	1 050

Source: SIPRI worksheets.

Appendix 1D

Table 1D1. The spread of sophisticated weapons to the third world: supersonic aircraft, 1955–1975

	1955	1956	1957	1958	1959	1960	1961	1962	1963	1964	1965	1966	1967	1968	1969	1970	1971	1972	1973	1974	1975
Israel	×	×	×	×	×	×	×	×	×	×	×	×	×	×	×	×	×	×	×	×	×
Taiwan				×	×	×	×	×	×	×	×	×	×	×	×	×	×	×	×	×	×
India				×	×	×	×	×	×	×	×	×	×	×	×	×	×	×	×	×	×
China					×	×	×	×	×	×	×	×	×	×	×	×	×	×	×	×	×
Cuba								×	×	×	×	×	×	×	×	×	×	×	×	×	×
Egypt								×	×	×	×	×	×	×	×	×	×	×	×	×	×
Pakistan								×	×	×	×	×	×	×	×	×	×	×	×	×	×
Iraq									×	×	×	×	×	×	×	×	×	×	×	×	×
South Africa									×	×	×	×	×	×	×	×	×	×	×	×	×
Indonesia										×	×	×	×	×	×	×	×	×	×	×	×
Algeria											×	×	×	×	×	×	×	×	×	×	×
Iran											×	×	×	×	×	×	×	×	×	×	×
Korea, North											×	×	×	×	×	×	×	×	×	×	×
Korea, South											×	×	×	×	×	×	×	×	×	×	×
Philippines											×	×	×	×	×	×	×	×	×	×	×
Afghanistan												×	×	×	×	×	×	×	×	×	×
Argentina												×	×	×	×	×	×	×	×	×	×
Ethiopia												×	×	×	×	×	×	×	×	×	×
Morocco												×	×	×	×	×	×	×	×	×	×
Saudi Arabia												×	×	×	×	×	×	×	×	×	×

Thailand	×	×	×	×	×	×	×	×	×	×
Viet-Nam, North	×	×	×	×	×	×	×	×	×	×
Lebanon		×	×	×	×	×	×	×	×	×
Syria		×	×	×	×	×	×	×	×	×
Viet-Nam, South		×	×	×	×	×	×	×	×	×
Kuwait			×	×	×	×	×	×	×	×
Libya			×	×	×	×	×	×	×	×
Paraguay			×	×	×	×	×	×	×	×
Jordan				×	×	×	×	×	×	×
Nigeria				×	×	×	×	×	×	×
Sudan					×	×	×	×	×	×
Brazil							×	×	×	×
Colombia							×	×	×	×
Abu Dhabi								×	×	×
Bangladesh								×	×	×
Singapore								×	×	×
Venezuela								×	×	×
Somalia									×	×
Tanzania									×	×
Uganda									×	×
Yemen									×	×
Zaire									×	×
Malaysia										×

Source: SIPRI data

Table 1D2. The spread of sophisticated weapons to the third world: long-range surface-to-air missiles, 1958–1975

	1958	1959	1960	1961	1962	1963	1964	1965	1966	1967	1968	1969	1970	1971	1972	1973	1974	1975
China	×	×	×	×	×	×	×	×	×	×	×	×	×	×	×	×	×	×
Taiwan		×	×	×	×	×	×	×	×	×	×	×	×	×	×	×	×	×
Cuba				×	×	×	×	×	×	×	×	×	×	×	×	×	×	×
Indonesia				×	×	×	×	×	×	×	×	×	×	×	×	×	×	×
Egypt						×	×	×	×	×	×	×	×	×	×	×	×	×
Iraq						×	×	×	×	×	×	×	×	×	×	×	×	×
Israel						×	×	×	×	×	×	×	×	×	×	×	×	×
India								×	×	×	×	×	×	×	×	×	×	×
Korea, South								×	×	×	×	×	×	×	×	×	×	×
Afghanistan									×	×	×	×	×	×	×	×	×	×
Algeria									×	×	×	×	×	×	×	×	×	×
Iran									×	×	×	×	×	×	×	×	×	×
Korea, North									×	×	×	×	×	×	×	×	×	×
Saudi Arabia									×	×	×	×	×	×	×	×	×	×
Viet-Nam, North									×	×	×	×	×	×	×	×	×	×
Syria										×	×	×	×	×	×	×	×	×
Thailand												×	×	×	×	×	×	×
Singapore														×	×	×	×	×
Sudan														×	×	×	×	×
Zambia														×	×	×	×	×
South Africa																×	×	×
Libya																	×	×
Pakistan																	×	×
Somalia																	×	×
Uganda																	×	×
Brazil																		×
Abu Dhabi																		×

Source: SIPRI data.

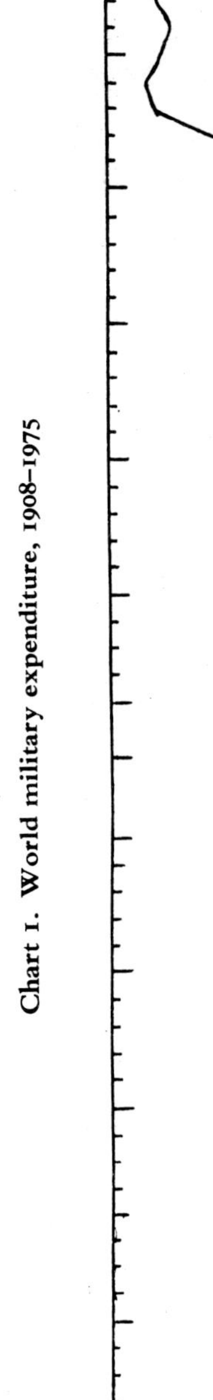

Chart 1. World military expenditure, 1908–1975

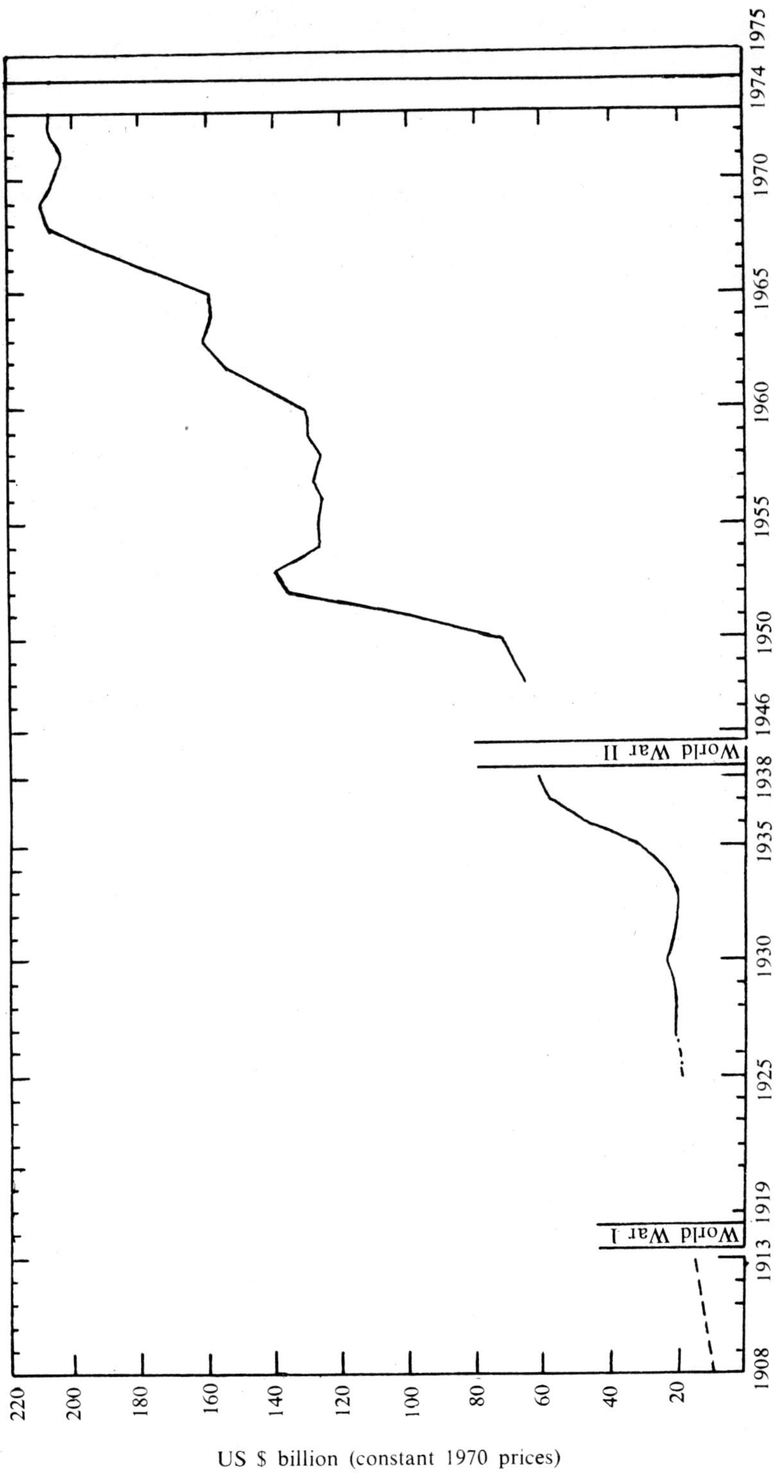

Chart 2. Indices of military expenditure in the United States and the Soviet Union, 1930–1974 (1970 = 100). Sources: SIPRI data.

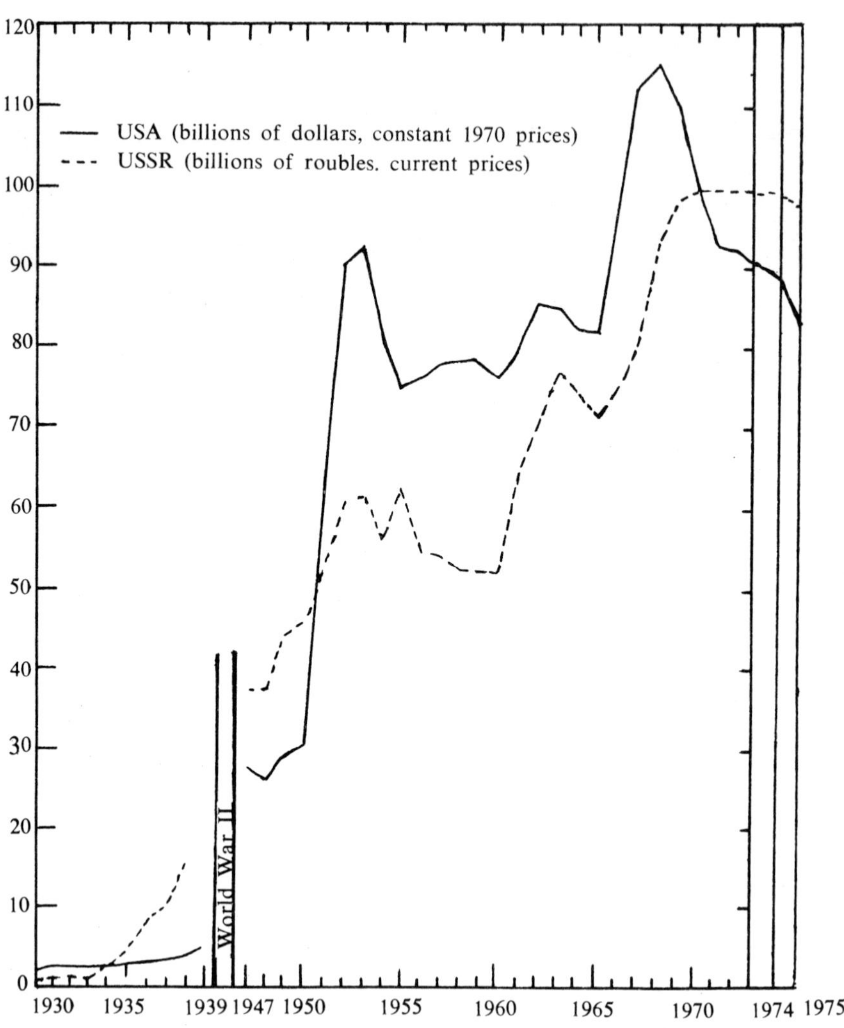

Chart 3. Value of major weapons transferred to the Third World: selected years 1950–1975

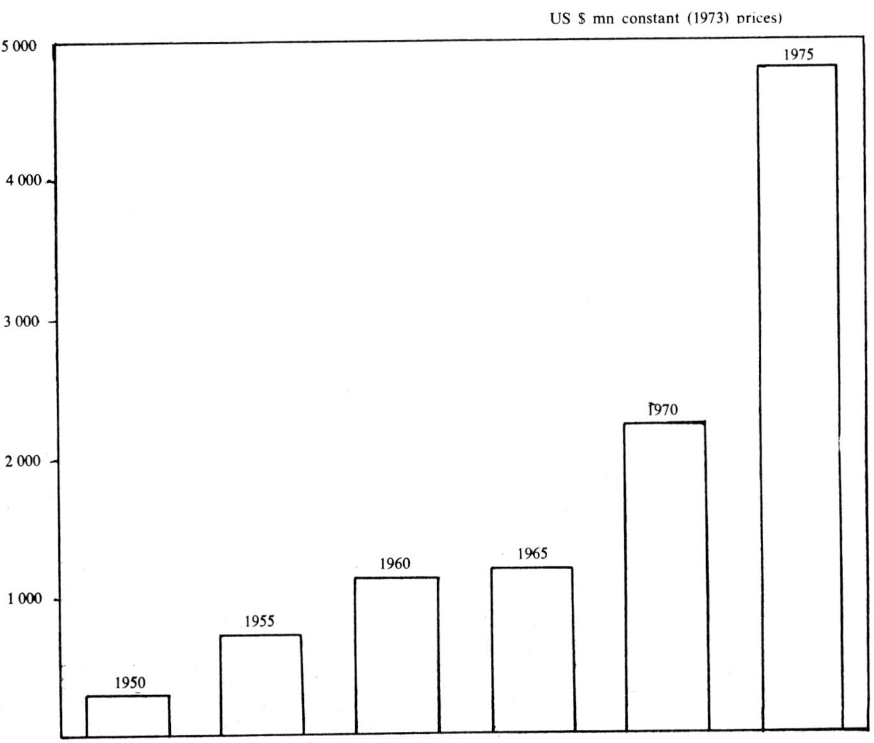

Chart 4. Military expenditure in the Middle East, 1960–1975

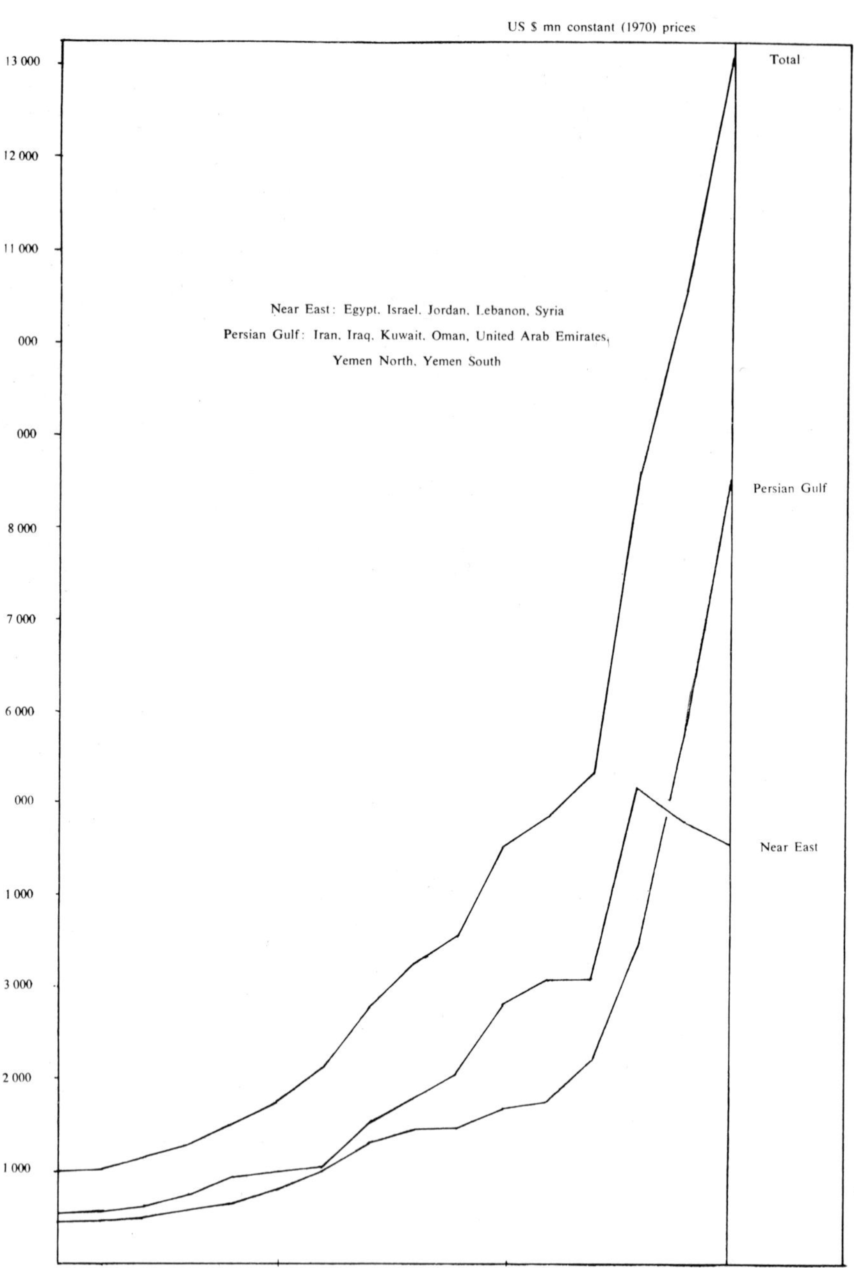

2. The nuclear momentum

1. *Thirty years of nuclear weaponry*[*]

On 3 July, 1944, the Danish physicist Niels Bohr addressed a memorandum to President Roosevelt and Prime Minister Churchill in which he said: 'The fact of immediate preponderance is that a weapon of unparalleled power is being created which will completely change all future conditions of warfare. Quite apart from the question of how soon the weapon will be ready for use and what role it may play in the present war, this situation raises a number of problems which call for most urgent attention. Unless, indeed, some agreement about the control of the use of the new active materials can be obtained in due time, any temporary advantage, however great, may be outweighed by a perpetual menace to human security.'

Bohr went on to plead for agreement among the then relevant great powers—the US, the USSR and the UK—on the control of nuclear energy *before* the completion of the atomic bomb and its use in war. He foresaw the conflict that would develop between East and West after World War II was over and the almost insurmountable difficulties of achieving any international control over nuclear developments in the resultant atmosphere of deep mistrust.

Both Roosevelt and Churchill talked with Bohr about his ideas, but unfortunately they were not impressed. One year after the Bohr memorandum was written—on 16 July, 1945—the world's first man-made nuclear explosion occurred at the Alamogordo, New Mexico, test site. Three weeks later, on 6 August, the second such explosion devastated the city of Hiroshima. And on 9 August, the third destroyed Nagasaki. The perception of 'a temporary advantage' from the use of nuclear weapons was allowed to outweigh the 'perpetual menace to human security' inherent in the existence of these weapons. No attempt was made first to establish a nuclear control system.

Today—30 years later—there are six powers (the US, the USSR, the UK, France, China, and India) which have exploded nuclear devices. And there is precious little evidence that any of them is prepared to give up the 'temporary advantage' it perceives in its nuclear status. What Bohr failed to appreciate was the overwhelming importance of 'temporary advantage' to the practising politician—no matter what his ideology may be.

---

* Shortened version of a paper by Frank Barnaby from *New Scientist*, Vol. 67, No. 961, 7 August 1975.

The technological race

In the 30 years since Hiroshima the nuclear arms race between the US and the USSR has been going on with almost as great a momentum as technological developments have allowed. Treaties designed to slow down this arms race have, in practice, had little effect on it. Considerable improvements have been made in nuclear-weapon design and in the methods of delivering nuclear warheads onto their targets.

The Hiroshima and Nagasaki bombs—euphemistically called Little Boy and Fat Man—were, on present standards, very primitive devices. Little Boy, which killed nearly 100 000 people, used about 60 kilograms of uranium-235 (about the size of a football) to obtain an explosion equivalent to that from 13 500 tons of TNT (13 kt)—an efficiency of only about one per cent. The cylindrically-shaped bomb, 3.2 metres long and 73 cm in diameter at its widest point, weighed about 4 500 kilograms. Fat Man was about one order of magnitude more efficient—about 8 kilograms of plutonium-239 (about the size of a grapefruit) were used to obtain an explosion of 22 kt. The egg-shaped bomb, 3.3 metres long including the fins and with a maximum diameter of 1.5 metres, weighed about 4 000 kilograms.

On 29 August, 1949, the Soviet Union exploded its first nuclear weapon, with a yield of about 20 kt, near Semipalatinsk in Central Asia. By this time, the Americans had exploded eight nuclear devices, with yields of up to 50 kt, and had stockpiled in their nuclear arsenal a few hundred nuclear weapons with a total yield of some 10 000 kt (10 megatons).

During the decade of the 1950's, an extraordinarily large number of major developments took place in nuclear weapon design and in nuclear delivery systems—a rate of innovation unparalleled before or since. It began in January 1950, when President Truman ordered the full-scale development of a thermonuclear weapon—a programme so successful that the first thermonuclear reaction was achieved in May 1951 at Eniwetok Atoll in the Pacific. According to Professor Herbert York, ex-Director of the Livermore Radiation Laboratory, the yield of this relatively crude device was about 200 kt. In November 1952, the first significant US thermonuclear (i.e. predominantly fusion) explosion took place on Elugelab Island, Eniwetok Atoll. The physical size of this device was enormous—a 6.6 metre long cylinder with a volume of about 500 cubic feet, weighing at least 200 tons. The yield of the explosion was about 10 mt. The Russians were not lagging very far behind the Americans—they exploded a thermonuclear device on 12 August, 1953. According to York, this had a yield of about 400 kt.

The early Soviet and American thermonuclear devices were not deliverable bombs. The first thermonuclear bomb was, in fact, set off by the Americans at Bikini Atoll on 1 March, 1954, giving a yield of 15 Mt. And a comparable Soviet bomb was exploded on 23 November, 1955, with a yield of about 2 Mt. The Americans have never set off a more powerful explosion than this 15 Mt one. The Russians, however, have exploded thermonuclear devices with yields of up to about 60 Mt—the biggest was exploded at

Novaya Zemlya in October 1961, probably more for political impact than because of military demand. It is of interest to note that the biggest pure fission bomb ever exploded was a 500 kt American device in November 1952.

Yield-to-weight ratio

A bomb's effectiveness is best measured by the ratio of its yield to its weight. Enormous increases in this ratio have been achieved. The yield-to-weight ratio of a conventional bomb in 1945 was no more than about 0.7 (it has not changed much since). For the Hiroshima bomb the ratio was about 3×10^3—an increase by a factor of 4×10^3. The largest-yield fission bomb probably had a yield-to-weight ratio of about 10^5. The ratio was increased in the mid-1950s to about 10^6 by the early thermonuclear weapons. Since then it has probably gone up by another factor of ten to a value which is only about eight times less than the theoretical maximum for fusion devices. A modern MIRV (multiple independently targetable re-entry vehicle) warhead, for example, would typically have a yield of about 200 kt and a weight of less than 100 kilograms. A Soviet SS-9 warhead may have a yield of 25 Mt and a weight of about 5 000 kilograms.

The physical dimensions of nuclear weapons have, of course, been much reduced. The Davy Crockett, for example, was a scaled-down version of Fat Man. The total length of the device, now withdrawn from US Army Service, was about 60 cm and its maximum diameter was about 30 cm. It had a yield of 0.25 kt. Nuclear artillery shells, often smaller versions of Little Boy, would be less than one metre in length and 15 cm in diameter. The US M-109 15-mm artillery shell has a yield of 2 kt and modern thermonuclear weapons would be of similar size.

Improving the design of nuclear weapons has required many test explosions. According to SIPRI figures, the US and the USSR have exploded about 1 000 nuclear devices in the past 30 years—on average, one every ten days.

Improved delivery systems

The first nuclear force—the US Strategic Air Command (SAC)—was formed in March 1946, seven months after the end of World War II, as a fleet of B-17 and B-29 medium bombers. In 1948, B-36 heavy long-range bombers and B-50 medium bombers began to replace the World War II vintage bombers in SAC. And, with bases in England and the Far East, SAC had by this time become an undeniably credible strategic threat to the USSR.

The SAC has continuously improved the performance of its bombers. High performance B-47 jet bombers were introduced in 1950, B-52 all-jet heavy long-range bombers in 1955, and FB-111 medium-range bombers in 1969. A new US strategic bomber, the B-1, is now being flight-tested—a decision on the procurement of this aircraft is to be made next year. In addition to free-fall thermonuclear bombs, SAC aircraft carry attack missiles with nuclear warheads.

In 1956, the USSR began deploying long-range bombers, the Tu-20 Bear and the Mya-4 Bison. The USSR, however, did not build up such a large strategic bomber force as did the US. At its peak, SAC was flying about 2 000 bombers. The number is now down to about 450. But the USSR probably keeps only about 150 long-range bombers in its strategic air force.

In 1954, the US began intensively developing long-range strategic missiles to carry nuclear warheads over intercontinental distances. Six missile programmes were initiated: one in 1954 to develop Atlas intercontinental ballistic missiles (ICBMs); three in 1955 to develop Titan ICBMs, and Jupiter and Thor intermediate-range ballistic missiles (IRBMs); one in 1956 to develop Polaris submarine-launched ballistic missiles (SLBMs); and one in 1957 to develop Minuteman ICBMs.

1957 was a watershed for the nuclear arms-race. In May the first US IRBM, a Jupiter, was successfully flown. In August the Soviet Union achieved the first significant flight with an ICBM. And on 4 October and 3 November the Soviet Union launched the first satellites, Sputniks I and II into orbit. The US followed on 31 January with Explorer I, but the American satellite was much lighter (14 kilograms) than the Soviet ones (83 kilograms for Sputnik I and 508 kilograms for Sputnik II). The balance was redressed in the following December when the US successfully flight-tested its first ICBM, an Atlas A. This missile entered service in 1960, as did the first US nuclear-powered strategic submarine (The George Washington) equipped with 16 Polaris SLBMs. The USSR began deploying ICBMs in 1961.

In excess of any conceivable need

The speed of developments in missile nuclear delivery systems came as a surprise to most observers. Also surprising was the rate of missile deployments. By 1967 the US had deployed 1 000 Minuteman and 54 Titan II ICBMs and 656 SLBMs; its land- and sea-based strategic missile forces have remained at these numbers ever since. But the USSR has actively continued to deploy ICBMs of various types since 1961 so that it now has about 1 600.

In 1964 the Soviet Union began deploying SLBMs, capable of being fired from a submerged position, on nuclear-powered submarines. And in the same year the US Polaris A-3 SLBM, equipped with three multiple re-entry vehicles (MRVs), entered service and the development of MIRV for Poseidon SLBMs was approved. MIRVs were, however, first deployed on land-based missiles—the US Minuteman III—in 1970. Each of these missiles carries three 170 kt warheads. In the following year the first Poseidon MIRVed SLBMs were deployed—16 missiles to each submarine, with each missile carrying between 10 and 14 nuclear (40 kt) warheads. And in 1973 the USSR began extensively flight testing MIRVs. Because MIRVs allow each missile to deliver nuclear warheads on to a number of separate targets, their deployment greatly increases the power of nuclear arsenals.

In the 30 years since Hiroshima, the nuclear arsenals of the US and the USSR have grown so large as to be grossly in excess of any conceivable need, political or military, of either power. The US has deployed about 9 000 nuclear warheads in strategic missiles (ICBMs and SLBMs) and bombers, and the USSR has deployed about 3 000. By 1985 these numbers may well have been increased to about 18 500 and 9 500 respectively, SALT II notwithstanding. In addition, each power has deployed or stockpiled tens of thousands of tactical weapons.

The quality of nuclear delivery systems and warheads is still being continuously improved. The US, for example, is working on the Trident SLBM, designed to carry manoeuvrable re-entry vehicles (MARVs) of very high accuracy. A new submarine will carry 24 of these missiles—the first is to be operational towards the end of 1978. The accuracy of strategic nuclear warheads is already astonishingly high. Some US ICBM warheads, for example, have a CEP (the radius of the circle centred on the target in which half of a large number of ICBM warheads fired at the target will fall) of about 250 metres. And significant further improvements in the accuracy of warheads is foreseeable with present technology.

But perhaps the greatest present concern is the danger of a new round of nuclear-weapon proliferation. By 1985 about 35 countries will be in a position to produce nuclear explosives as by-products of peaceful nuclear programmes. How many nuclear-weapon powers will there be on the 40th anniversary of Hiroshima?

2. *Nuclear weapons: the ultimate absurdity**

If you were to ask the leaders of the nuclear powers why they feel it necessary to possess nuclear weapons, they would answer 'to ensure our national security'. They might also add that nuclear weapons help to stabilize international relations and prevent the outbreak of another general war. And if we review the details of the history of the nuclear arms race, we find that each nuclear power originally initiated its nuclear development programme for what they considered to be essential national security reasons. The United States began its programme in 1941 because it feared the possibility that the Germans might do so first. The Russians commonly said at the time that they initiated their high priority programme in 1945 in order to eliminate American 'nuclear blackmail'. The French started their nuclear weapon programme in the 1950's because they feared they could not rely on American nuclear strength in a crisis, and the Chinese initiated their programme

*Paper by Herbert York from *Bulletin of Peace Proposals*, Vol. 7, No. 1, 1976.

in order to provide some counterbalance to both the Soviet and the American nuclear forces.

But has the introduction of nuclear weapons really enhanced the national security of those who possess them? And is the arms race leading to international stability?

National security

In order to answer the first question, I shall turn to the specific case of the United States. I do so, not because it is really different from the other cases in any fundamental way, but because I personally understand it better.

In the late forties, following the Second World War, the United States was invulnerable to an attack by any other state. Our geographic situation and the strength of our military forces, coupled with our nuclear monopoly, made it impossible for any other nation to mount an effective attack on our homeland, using only conventional weapons. In 1949, the Soviets broke the American monopoly on nuclear explosives. By the mid fifties, they had sufficient nuclear weapons, and sufficient long-range bombers to enable them to mount an attack on the United States that could produce in the span of a single day a few million, perhaps a few tens of millions, casualties. For the first time in more than a century, our country had become seriously vulnerable to an attack by another. We reacted by developing several new varieties of weapons, including tactical nuclear weapons, long range rockets and, perhaps most importantly, the extremely powerful hydrogen bomb. At more or less the same time, the Soviets also developed such weapons. As a result, by the late sixties, the Soviets had acquired an attack force which could, for all practical purposes, kill our entire urban population in the span of less than an hour. We spent many billions of dollars trying to develop methods which could either prevent such an attack or cope effectively with its effects, but we were unable to do so. That situation has remained the same to the present time. We are absolutely incapable of preventing the Soviet rocket forces from completely destroying our country at this very moment. Of course, the reverse is also true. We could completely destroy the cities of the Soviet Union, and they do not possess any means for preventing us from doing so. At the present time, this capability for mutual destruction does not depend on who strikes first. As long as that last condition holds true, the situation has what the defence intellectuals calls 'crisis stability'. In other words, as long as the leaders of each country remain both rational and reasonably humane, each side, under the current circumstance, will be 'deterred' from attacking the other. Because nearly the entire population of both countries is at risk, that is, because we each possess what is often called an 'overkill' capability, the number of lives at risk cannot be changed very much in either direction by any new moves in the arms race. To summarize, as far as the vulnerability of the United States is concerned, the first twenty years of the nuclear arms race changed that country from an invulnerable nation to a nation that could be completely destroyed in a fraction of a day.

International stability

Let us now turn to the question of stability. In doing so, we will find it useful to sub-divide the question into two parts. The first part is the question of the stability of the Soviet-American relationship, neglecting, as we so often do, everyone else. The second part is the question of stability in the real world with its more than one hundred independent nations.

In the case of the strictly bipolar Soviet-American relationship, we must first note that it has so far been stable; that is, there has been no Soviet-American nuclear war, and it is obvious that the leaders of each country will go to very great lengths to avoid one. But this stability, even neglecting the rest of the world, is a fragile thing. It depends absolutely on the good sense and sanity of the leaders, and it requires that there never be any nuclear mistakes or accidents of any kind, technological, military, or administrative. Many analysts, and high government officials as well, have expressed their doubts about the ability of this current situation of stability to last indefinitely into the future.

In addition, and still considering the Soviet-American relationship only, qualitative technological changes could upset the current stable balance. The Strategy of Deterrence by means of 'Mutual Assured Destruction' requires that the parties involved possess nuclear forces which not only are essentially equal in size to each other, but which are also safe against the possibility of first strike. That is, the forces of each side must be protected and deployed in such a way that the number of weapons which would survive a surprise pre-emptive strike would still be great enough to achieve an obviously unacceptable level of retaliation (or revenge) on the attacker. At present, the forces of the USA and the USSR both satisfy this requirement, but this could change. For example, the simultaneous (or near simultaneous) introduction of multiple war heads, improved accuracy, and better means of detecting and destroying submarines, might make a first strike possible. This is, it might become possible for the side which struck first to destroy nearly all of the other side's weapons, and thus reduce the nuclear revenge it would undergo to an 'acceptable' level. To make matters worse, it is not necessary for a first strike possibility to really exist; if the leaders simply believe it exists, then the situation may become dangerously unstable. Of course, it may not be possible to achieve such a combination of technological capabilities, but we should all realize that tens of thousands of scientists and engineers on both sides are trying to achieve them, and we must bear in mind that such people have achieved some very remarkable (and even improbable) results in recent years.

Bipolar and multipolar arms races

The foregoing applies to a strictly bipolar Soviet-American arms race, and ignores the rest of the world. How does this analysis change if we consider these others? It seems to me to be completely obvious that the picture becomes much worse. There are now six nations that have tested nuclear explosives. Several others possess the technological base for doing so soon, and some of these may even have constructed weapons which they have so far avoided testing, in order to avoid stimulating their neighbours. Such nations are referred to as the 'near nuclear' countries. In addition, the steady and seemingly inevitable spread of nuclear power throughout the world will put nuclear weapons within reach of many more nations before the turn of the century. We can expect the pressures on some of these nations to 'go nuclear' to become irresistible. Several of them are in geo-political circumstances such that they feel themselves to be in very grave danger, and the bad examples furnished by the five nations now possessing large numbers of nuclear weapons will overcome any remaining doubts they may have. Simply because there will be so many 'hands on the trigger' the present near stability is bound to break down. More fundamentally, it will simply be impossible to maintain the kind of situation of parity or balance that now exists in the Soviet-American relationship. It will be inevitable that some nations will have relatively many bombs, and some will have relatively few. Some will possess sophisticated bombs with high accuracy, and some will have crude bombs. Some nations will have their weapons mounted on missiles which are hidden away in silos so that some of them will be sure of surviving a surprise attack, and some nations will have their bombs deployed on aircraft, which will be parked in the open and easily destroyed. There will therefore arise many situations in which a successful pre-emptive strike will either be possible or at least seem to be possible, and in which, therefore, neither the technical nor the political bases of the current bipolar stability will exist. In such a world the possibility of 'false alarms', and of nuclear accidents of all kinds will increase greatly. In such a world, stable nuclear 'deterrence' as we have known it will become impossible, and war will become inevitable.

A steadily worsening situation

In sum, the situation is becoming increasingly absurd. In each individual case, the nuclear states (current and future) develop and accumulate nuclear weapons in order to enhance their national security. In the past, the net result has been that the consequences of a nuclear war have become much more terrible as time went on, and in the future the probability of a nuclear holocaust occurring will steadily increase. This absurd situation can easily result in what we may call the ultimate absurdity. That is, it may result in a situation such that we will simultaneously eliminate both ourselves and all of our other social and political problems.

It seems clear that this problem must be solved, no matter how difficult or inconvenient the means for solving it may be. It also seems clear that the current attempts for solving it are entirely inadequate. Despite the efforts of both the United Nations Committee on Disarmament and the bilateral Strategic Arms Limitation Conference, the situation has clearly grown steadily worse. The number of nuclear warheads is steadily increasing, the number of nuclear states has increased and threatens to accelerate, and a steady stream of qualitative improvements threaten to destroy such stability as even the current so called 'balance of terror' may possess. In order to reverse all of these trends, much more extreme measures than those now employed will be needed. As a minimum, control of at least the most sensitive nuclear operations will have to be given over to an international agency, and the present curtain of secrecy surrounding the nuclear world will have to be very largely removed. Such steps will require real inroads into national sovereignty and real freedom in the flow of both people and information across now closed (or tightly controlled) frontiers. These changes may seem impossible, and maybe they really are. Even so, we must face them and try to solve them, because the anarchistic multipolar nuclear world we are now rushing straight into, is even more impossible.

3. *Counterforce: new round in the nuclear arms race**

As time goes on, the severe shortcomings of nuclear deterrent doctrines are becoming widely appreciated and these doctrines are becoming known for what they are—inhumane, irrational, positively dangerous and a bar to progress in disarmament.

Until 1974, US and Soviet policies concerning strategic nuclear weapons relied primarily on the doctrine of 'mutual assured destruction'. The main feature of this deterrent doctrine is a certain ability to inflict massive destruction on the enemy population and industry in a retaliatory attack, following a massive nuclear strike by the enemy.

In 1973, a number of articles and statements appeared in the United States calling for a revision in US strategic weapon policy. The essence of the change being suggested was to give much more emphasis than hitherto to the 'counterforce' capabilities of the nuclear forces. Counterforce strategy does not replace deterrence: rather it supplements it with the additional capability to strike, either pre-emptively or in response to an attack, at the opponent's military targets, including hardened missile silos. Such a strategy requires a large number of accurate, powerful nuclear warheads targeted not against cities and industrial and transportation centres but against military installations. In addition, a counterforce strategy implies the capability of fighting a nuclear war if deterrence fails to prevent its outbreak. US Secretary of Defense Schlesinger announced on 10 January 1974 that the USA actually intends to adopt a counterforce strategy as a strategic nuclear option and, to this end, is improving the accuracy of delivery of its nuclear weapon systems.

* From *SIPRI Yearbooks*, 1974 and 1975.

Counterforce capability

In order to assess the import of this announcement it is essential to observe, first, that in reality the United States has by now already had a counterforce capability and targeting policy for its strategic missiles for several years. A comparison between the number of Soviet urban centres worth a nuclear attack and the number and characteristics of nuclear warheads in the US strategic arsenals leads to the unavoidable conclusion that most of the US strategic nuclear weapons are targeted against military targets in the Soviet Union. The Soviet Union has at most 100 urban and industrial centres that would be targeted with nuclear warheads. Clearly even if the number of targets in the Soviet Union is not 100 but twice that, and even if one assumes an equal number of targets in the People's Republic of China, all the civilian targets do not exceed 400. Furthermore, even if we assume that each of these targets is double-targeted to allow for possible missile failure, the number of warheads aimed against them does not exceed 800. Not all the warheads in the US arsenal are, of course, available for launch at all times. The total number of warheads ready to be launched at any instant is 4 500 on missiles and probably no less than 800 or so on B-52s. Since the number of warheads needed to destroy securely all the significant civilian targets in the Soviet Union and China is 800, over 4 000 nuclear warheads must have been targeted on military targets for the past several years. Since most of the US missiles have multiple-target storage memories and performance indicators, the strategic force has considerable flexibility, recently augmented by the introduction of the command data-buffer system*. Therefore, in effect, the USA already possesses a counterforce nuclear arsenal which includes a variety of warheads and is controlled by a flexible command system.

Rationalizing new strategic weapon systems

The counterforce strategy, as enunciated by Schlesinger, includes the humanitarian implication that the USA should acquire a 'surgical strike' capability, that is, the ability to destroy a military target near or in an urban centre incisively without adverse effects to the civilian population. This posture, while acceptable to the public, requires warheads with very high accuracy and very small nuclear charge; such weapons are not currently deployed in the US arsenal. But mini-nuclear warheads and terminal guidance that endows warheads with pinpoint accuracy are two weapon developments that have reached a state of completion and are ready for procurement and deployment. The nuclear weapon laboratories of the US government have perfected small warheads in the low- or even fractional-kiloton region, and are pressing for their adoption in the US arsenal. Project

* The command data-buffer system will enable an ICBM to be switched to a new target in a relatively short time (about 20 minutes). Without the system the process, which then entails reprogramming each missile's computer, would require up to 36 hours to complete.

ABRES (Advanced Ballistic Re-Entry Systems), managed by the Space and Missiles Systems Organization (SAMSO) of the US Air Force, has succeeded in perfecting a terminal guidance system that, when incorporated in a manoeuvrable re-entry vehicle (MARV), can guide it onto a prescribed target. Neither terminal guidance on MARV nor mini-nuclear warheads can be justified in the framework of an assured destruction posture; they can, however, be presented as essential weapons in a strategy of counterforce.

The introduction of MARV will eventually justify the development of yet another strategic weapon system that could not be justified by the strategy of deterrence, namely, the mobile land-based ICBM. It is almost certain that nuclear warheads on MARV will have the yield and accuracy to assure destruction of ICBMs in reinforced silos. It is then quite probable that, following procurement of a US MARV and invoking a projected acquisition of the same type by the Soviet Union, the US Air Force will request funds to deploy mobile ICBMs as a method of protecting the land-based component of US strategic forces.

Serious ramifications

The formal adoption of the counterforce posture by the USA will tend to induce a number of serious ramifications.

1. By insisting on the capability for a 'surgical strike' it tends to move strategic nuclear weapons closer to the tactical arena, and to blur further the distinction between tactical and strategic nuclear strikes. The elimination of this 'firebreak', that has so far prevented the use of nuclear weapons in tactical operations, has been fervently pursued by the military in the past quarter century and adamantly opposed by those who believe that use of even small nuclear weapons for a 'surgical strike' will inevitably lead to escalation and nuclear holocaust.

2. The introduction of MARV will provide the USA with undeniable first-strike capability against land-based missiles of other countries. For the foreseeable future MARV will introduce an intense crisis instability by making Soviet land-based missiles vulnerable to a first strike.

3. It is almost certain that MARV will be used as an argument by the Ballistic Systems Division in the USA and the Strategic Rockets Division in the Soviet Union in favour of deploying mobile land-based ICBMs.

But perhaps the greatest danger implicit in the posture of counterforce is the effort to make the 'surgical strike' credible and acceptable. The fallacy of the claim that a nuclear weapon, no matter how accurate and small, can be used in or near an urban complex—where most of the militarily interesting targets actually lie—without detrimental effects to the civilian population has been shown in several recent studies. Aside from the immediate blast effects, higher order effects and radiation will certainly take a heavy civilian toll whether the missile is actually targeted against an urban centre or an airfield a few miles away from it. The establishment of the 'surgical strike' as an option available to the military appears to be a reckless and unnecessary escalation of the dangers inherent in the nuclear confrontation between the United States and the Soviet Union.

Deterrence and the risks of war

It is doubtful, at least in the long term, that nuclear weapons have had the political utility claimed for them, since their existence has indeed greatly decreased world security and most particularly the security of those countries actually having nuclear weapons.

No nuclear strategy, past or present, based on deterrence is credible to rational persons, particularly when the powers concerned are equally armed. Such strategies must be seen for what they are—mere rationalizations for the development and deployment of the weapons made available by military technology.

The commonly held view that the very destructiveness of nuclear weapons precludes the outbreak of thermonuclear war is incorrect. Even if 'rational behaviour' is assumed, thermonuclear war is unlikely to occur only if it is believed that neither side can win. If one power perceives a chance of winning, then there is a risk that it will decide to strike while it has the advantage. Moreover, in the event of a serious crisis, the side placed at a disadvantage may, if it believes a nuclear war inevitable, attack first in the hope of reducing the damage it thinks it is bound to suffer.

At present, and for the foreseeable future, a general nuclear war could not be 'won' by either side. Because of the massive deployments of nuclear weapons of all types by both sides, the likelihood of either side escaping unscathed following an initial attack is virtually nil. But no one knows whether the present technological situation will continue for decades. There are various courses of development, some of which are now being pursued by both the USA and the USSR, which tend toward the acquisition of a first-strike capability. These are likely to produce periods of extreme instability and heightened risk of nuclear war.

Nuclear forces are also maintained on the grounds of deterring incursions or acts of political or economic blackmail. The requirement that the populations of not only the USA and the USSR but also other countries be placed in jeopardy of nuclear war cannot be justified by the potential deterrence of lesser conflicts. In addition to the possible erosion of the more 'rational' reasons for the non-use of nuclear weapons, there is the great and ever present danger that nuclear war will come about by accident, miscalculation or madness. No nuclear strategy can deal effectively with this possibility.

Making nuclear war more thinkable

President Nixon said in his State of the Union message: 'We must never allow America to become the second strongest nation in the world'. This type of statement from the political leadership encourages those groups within societies which press for a continuation of the arms race and for maximum effort in military research to develop weapons to fight and 'win' a nuclear war. Counterforce strategy can be seen as yet another step in this direction. In making nuclear war more 'flexible', it makes it more thinkable, more tolerable and consequently more probable.

Because it cannot be said that efforts to acquire a first-strike capability will not eventually succeed, albeit in the long term, the only credible policy to reduce the probability of nuclear war to an acceptable level is far-reaching nuclear disarmament as a first step in a planned programme leading to general and complete disarmament. But perhaps the most immediate requirement is for the political decision-makers to understand and to recognize the urgent need for nuclear disarmament so that they will resist those vested interests within their societies which press for the application of all possible technological advances to weapon system development, for the deployment of all new weapons that are developed and for the maintenance and improvement of all existing weapon systems.

The inherent dilemma

The debate on nuclear deterrent policies underlines the inherent dilemma produced by the existence of large nuclear arsenals. All nuclear doctrines must have severe shortcomings, mainly because they cannot reduce the probability of the use of the weapons to an acceptably low level.

By continuing to improve both the quality of nuclear weapons and the numbers of these weapons, the USA and the Soviet Union imply that they perceive a military and/or political utility in them. The official statements of the two powers also imply that they believe that not only does nuclear deterrence work but that it is a good policy. How then can they expect other powers not to acquire nuclear weapons? If nuclear deterrence works in Europe, it should also work in Asia. The USA and the Soviet Union cannot contribute to the prevention of nuclear weapon proliferation unless they show by their actions that they see no utility in nuclear weapons. The only way in which they can demonstrate this, is to reduce their nuclear arsenals, as a step to the total abolition of nuclear weapons.

4. *Offensive missiles**

The performance characteristics of nuclear warheads, and the ballistic missiles that transport them to their targets, are rarely discussed in any detail in public. This is often attributed to the secrecy that, of necessity, surrounds these weapons, or to the technical complexity of the subject, although there is enough information, publicly available and relatively straightforward in

* From *Offensive Missiles*, SIPRI, 1974.

nature, to form the basis for a public debate on this topic. It is also often argued that intricate technical details of the military capabilities of these weapons are not relevant in a public debate concerning the acquisition and deployment of new weapons, because such factors as political perceptions or appearances, psychological pressures, or bureaucratic expedience may be more influential than technical details in deciding the outcome of the debate. The public and the politicians, it is frequently argued, understand simple concepts such as the numbers of missiles various countries possess and the physical size of the rockets, but not such intangibles as accuracy or reliability. Yet it is these latter performance parameters that are actually important and not the numbers and sizes of launchers.

In the public domain, a comparison of the US and Soviet nuclear arsenals, based as it is on the number and size of missiles, favours the Soviet Union; therefore it justifies the recent requests for new and expensive US programmes to replace the present nuclear weapons with new ones. But an examination of the two arsenals based on parameters relevant to a strategy of counterforce reveals that the United States has always had a considerable margin of superiority and that this superiority will continue, even if no new programmes are initiated. Furthermore, it shows that if the new programmes are pursued to their completion they will result in at least two unfavourable developments: nuclear war will become more probable, and the Soviet military will find justification to demand from their political leadership new missile-construction programmes beyond the present one that entails replacement of the second-generation missiles with MIRVed ones. Such an event will virtually guarantee that the accumulation of increasingly sophisticated missiles on both sides will continue.

Effects of a nuclear weapon

A land-based intercontinental ballistic missile consists of a rocket, usually a multi-stage arrangement, propelled by a solid-fuel motor and guided by an inertial guidance system. It carries either a single re-entry vehicle containing a nuclear explosive warhead, or in the case of MIRV, several re-entry vehicles, each containing a nuclear warhead and capable of being independently targeted. Each warhead may weigh as little as a few tens of kilograms since the nuclear charge itself may weigh as little as 10–20 kilograms.

The destructive effect of a nuclear weapon results from the almost instantaneous release of enormous amounts of energy: the explosion of a one-megaton warhead releases, within a few nanoseconds (10^{-9} second), energy equivalent to that released by the explosion of one million tons of TNT, that is, about 10^{15} calories, generating, in the immediate vicinity of the explosion, temperatures of millions of degrees centigrade. This enormous heat causes the sudden expansion of the air around the point of explosion, which in turn gives rise to a shockwave in which pressures reach 10^5 pounds per square inch (about 7 000 atmospheres), and then decay rapidly as the shockwave propagates outward from the point of explosion.

Although the same nuclear warhead can be used against a city or industrial complex (countervalue attack) or against a reinforced concrete silo housing a missile (counterforce attack), the performance characteristics of the warhead and the missile that are relevant to each type of attack are very different.

Shockwave overpressures (pressures above atmospheric) of 20 pounds per square inch will usually kill an unprotected human being, while overpressures of five pounds per square inch will demolish an ordinary brick house. Since overpressure is proportional to the energy released by a nuclear charge and inversely proportional to the cube of the distance from the point of explosion, the larger the energy yield of the weapon the further from its point of impact will be the perimeter of total destruction. A one-megaton nuclear weapon, for example, will create a five-pounds-per-square inch overpressure four kilometres from the point of explosion, and will thus destroy all houses in an area 50 square kilometres around its point of impact. Therefore, a missile aimed at a city (a countervalue attack) does not have to be very accurate, since the destruction of property and life caused by the weapon will be immense no matter exactly where in the city it lands.

Although blast effects can prove fatal to people, by far the most lethal effect of a nuclear explosion is the thermal radiation released: the fireball created by a nuclear blast attains temperatures of tens of millions of degrees centigrade and therefore radiates very much like the sun. So nuclear weapons, unlike conventional weapons, destroy both buildings and human beings by direct thermal effects and by the fires and firestorm induced by the heat released in the explosion. Both the direct and induced thermal effects, whose magnitudes are proportional to the size of the warhead, extend to great distances from the point of explosion: the heat released by a one-megaton nuclear explosion, for example, will cause paper to ignite 14 km away. An idea of the destructiveness of these thermal effects can be gained from the fact that about 50 per cent of the fatalities caused by the Hiroshima bomb (whose yield was about 15 kilotons (15 000 tons of TNT equivalent)) were due to primary or induced thermal effects, 30 per cent to nuclear radiation and 20 per cent to blast. Thermal radiation effects of a nuclear explosion, like blast effects, spread over very large areas and consequently a nuclear weapon can cause extensive damage to a city by fire, even if it is not delivered accurately against it.

An attack against urban or industrial centres, therefore, requires the delivery of large amounts of thermal energy and the creation of modest overpressures (five to 10 pounds per square inch) over very large areas. This can be achieved either by delivering a high-yield nuclear weapon somewhere in the area of the target or by scattering several small weapons, even at random, over the area. These are the performance requirements of a countervalue nuclear-weapon system.

Counterforce performance requirements

The performance requirements of a weapon destined for counterforce use, that is, for destroying a missile inside a reinforced concrete silo, are quite different: the weapon must deliver a very large amount of energy against a very small target, and must therefore be very accurate. Overpressure decreases rapidly with distance from the point of impact of the nuclear weapon; therefore, since a silo is designed to withstand overpressures of hundreds of pounds per square inch, and is immune to thermal effects, it will remain intact unless the weapon lands in its immediate vicinity. Consequently, the ability of a warhead to destroy a silo depends strongly on the accuracy with which it is delivered to its target. As a matter of fact, the lethality of a warhead against a silo rises much more rapidly with improvements in accuracy than with increases in yield. For example, a weapon with a yield 10 times greater than a Minuteman III warhead is, given the same accuracy, five times more effective in destroying a silo, but a weapon that has the same yield as a Minuteman III warhead, but which is 10 times more accurate, is 100 times more lethal to a silo. So it is the accuracy of the missile that delivers the warhead that is relevant in a counterforce nuclear arsenal, and not the size of the missile.

Electromagnetic pulse and 'fratricide'

Another phenomenon associated with a nuclear explosion that is relevant for a counterforce weapon is the so-called Electromagnetic Pulse (EMP). A nuclear explosion ionizes the atoms of the atmosphere in its immediate vicinity, giving rise to distributions of negative and positive charges. These distributions are not spherically symmetrical in space and therefore create oscillating electric and magnetic fields that, near the point of detonation, can reach strengths of tens of kilovolts per metre and several hundred gauss, respectively. Such fields can penetrate an unshielded silo and destroy the complex and delicate electronic equipment of a missile and of its launching facilities, even if the silo can withstand the overpressure created by the blast.

The EMP, together with the violent movement of the air mass near the explosion, the large amounts of debris that the explosion causes to rise rapidly into the upper atmosphere, and the persistently high level of radioactivity emanating from the expanding fireball, combine to create another effect, known as 'interference' or 'screening' or 'fratricide' (the latter term was coined by the Pentagon). 'Interference' refers to the fact that these results of a nuclear explosion make it difficult, if not impossible, to deliver a second re-entry vehicle to the same point, soon after the first one has arrived and detonated. As the second re-entry vehicle enters the atmosphere near the point where the first exploded, it encounters high densities of dust that can cause its ablative shield to burn prematurely, or it can be deflected off-target by the violent winds that persist in the area for considerable periods of time, or, if it arrives a few seconds after the first weapon, it could even be destroyed by the EMP or the nuclear radiation emanating from the rising fireball. Thus, the interference effect does not physically forbid the use of several re-entry vehicles against the same silo, but renders such a targeting schedule inefficient and uncertain.

The uncertainties introduced by these two effects, and their interaction, are such that a missile force cannot be considered to possess unmistakable countersilo capabilities unless it includes weapons with yield and accuracy high enough to destroy a silo with a *single* re-entry vehicle. Even if a MIRVed missile has, in principle, enough warheads with enough accuracy and yield to destroy a silo with certainty, it is doubtful that it has this capability in real life. Neither can a prudent strategist use such a weapon in a first strike against an opponent's silos: the interference effect makes the efficacy of all but the first re-entry vehicle that reaches the silo doubtful, since it would probably cause the destruction, or the wandering off-target of all subsequent ones. Therefore the interference effect must be always borne in mind in calculating the countersilo-attack capability of a country that does not possess missiles accurate enough to destroy a silo with a single warhead.

US programmes for missile improvements

At the present time, neither the Soviet Union nor the United States has the capability to destroy, using ballistic missiles alone, the land-based ICBM force of the other country with any reasonable probability. The tangible superiority of the US land-based missiles, both in the numbers of independently-deliverable warheads (N), and in the total countersilo kill capability ($K \cdot N$), leaves the two new sets of programmes for the drastic improvement of the US missiles, proposed by Mr Schlesinger, without rational justification.

The first of the US strategic-missile improvement programmes that can be implemented with existing technology will result in a twofold increase in the yield of the Minuteman III and SLBM warheads, and an improvement in the accuracy (*CEP*) of the Minuteman III re-entry vehicles to about 250 metres (0.13 nautical miles) and of the missiles intended for the Poseidon and Trident submarines to somewhat better than 0.2 nautical miles.

The second programme proposed by Mr Schlesinger aims at introducing terminal guidance on US re-entry vehicles, endowing them with accuracies of 30–50 metres. Although the techniques required for terminal guidance of manoeuvrable re-entry vehicles (MARV) are beyond the research stage, they still require extensive development before they can be incorporated into reliable weapons systems, and so it may be 10 more years before these weapons enter the US strategic arsenal.

Threat to arms limitation efforts

Yet such staggering lethality does not offer any practical superiority, because even if all the Soviet land-based missiles are destroyed, there will still remain in the Soviet arsenal several hundred submarine-based missiles that can completely devastate the US population and industrial centres. The fact remains that a few dozen missiles can deter even the most daring political leader from launching a nuclear attack. The military leaders of the Soviet

Union may be lured into responding to these US programmes with similar efforts to upgrade the accuracy of their missiles, and thereby offer justification to the technobureaucracy of the US defence establishment for still further programmes. But independently of the precipitousness with which the Soviet Union responds, two predictions can be safely offered:

(1) Since the properties of matter do not allow accuracies better than 30 metres or so (neither are they necessary), Soviet missile accuracies will sooner or later catch up those of US missiles, restoring, sometime in the early 1990's, the parity of forces that will have existed twice before—in the early 1970's in number of launchers and in the early 1980's in number of independently-deliverable re-entry vehicles. Thus Nature will terminate the competition for increased lethality of nuclear weapons that political leaders seem unable, and the military unwilling, to end.

(2) If the Soviet leaders decide that the survivability of their strategic rocket forces warrants additional expense, they can replace their ICBMs with mobile ones, thereby rendering US improvements in accuracy useless, but at the same time complicating further the complex task of verification by national means of inspection of the number of missiles each country possesses. Terminal guidance presents a threat not to the strategic missiles of the USSR, but to arms-limitation efforts.

5. *The nuclear 'balance of terror' in Europe**

At present, the military forces of the NATO countries are armed with almost 10 000 nuclear weapons. Most of them (7 000 or so) were brought to Europe by the Americans, but the British and French own substantial numbers of such weapons also. The power of these weapons varies widely from one type to another, but all are enormously murderous and destructive when compared to ordinary explosives. The smallest have an explosive power equivalent to about 1 000 tonnes of chemical explosives (1 tonne= 1 000 kg; 1 000 tonnes= 1 kt, or 1 kilotonne) and the largest have more than 1 000 000 tonnes equivalent (= 1 Mt, or 1 megatonne).

NATO authorities commonly state that the main purpose for possessing all these nuclear weapons is to 'deter' an attack on Western Europe. NATO authorities do, of course, recognize that deterrence might somehow fail. Since Western strategists commonly contend that Europe cannot be defended without using nuclear weapons, they have therefore developed war plans which call for the actual use of nuclear weapons in the event that deterrence fails.

The Soviet forces deployed in and facing Europe are also equipped with nuclear weapons specifically designed for so-called 'tactical' or 'theatre' uses. The number of such weapons is commonly said to be about one-half the

* Shortened version of a paper by Herbert York from *Ambio*, Vol. 4, No. 5–6, 1975.

number of American bombs in NATO Europe, but the power of the Soviet A-bombs is on the average higher than those deployed in the West. The Soviets for years now have said they have no intention of attacking the West, but they add that if for any reason NATO should use its nuclear weapons against them, they will reply in kind.

Soviet strategists and commentators generally belittle the notion that a nuclear war can be kept limited, and they show no signs of being interested in any restraints on precisely how nuclear weapons may be used once their use has been initiated. Similarly, currently available Soviet writings on nuclear strategy and Soviet training procedures and documents give no indication of their being interested in either reducing or enhancing collateral damage to civilians. They seem to believe that any nuclear war will be destructive to a degree far beyond anything known in the past, and they appear to be determined to have most of that destruction happen to somebody else.

The answer to the question of *whether* a nuclear war will take place in Europe is determined at least as much by NATO deployments and doctrines as by Soviet deployments and doctrines. However, the answer to the question of *what* will happen to Western Europe when a nuclear war takes place, will be almost entirely determined by what the Soviets actually do with the nuclear weapons in their possession. Therefore, any discussion of the results of nuclear war in Europe must focus mainly on what weapons the Soviets have, how they might be used, and what could result from such use.

Soviet nuclear weapons

The number of Soviet tactical nuclear weapons deployed and programmed for use in and against Europe is usually given as about 3 500. Some details of the various types of tactical delivery systems, including estimates of numbers, ranges, and explosive power, are presented in Table 2.1.

Table 2.1. Soviet tactical and strategic nuclear systems

Type	*Number*	*Range*	*Yield***
A. Tactical systems			
1. Howitzer	100	25 km	few kt
2. Short-range, land-based missiles	1 000	6–800 km	10 kt–500 kt
3. Medium-range, land-based missiles	600	1 500–3 000 km	0.5 Mt–3.0 Mt
4. Short-range, sea-based missiles	500	100–500 km	20 kt–500 kt
5. Nuclear-capable tactical aircraft	2 400	600–3 000 km*	kilotonnes to megatonnes
B. Strategic systems			
6. Strategic missiles, land- and sea-based	2 400	to 12 000 km	Mt (to 25 Mt)
7. Strategic aircraft	150	10 000 km (?)	Mt (to 25 Mt)

Notes: * strike radius
** 1 kt = 1 kilotonne
1 Mt = 1 megatonne

Source: Adapted from Jeffrey Record, *US Nuclear Weapons in Europe* (Brookings Institution, Washington, DC, 1974).

I shall from here on focus exclusively on Soviet weapons of type 3, that is the 600 medium-range ballistic missiles with megatonne warheads. The justification for excluding all other Soviet weapons is that unlike types 4 through 6, these weapons have no other purpose than the bombardment of Europe and also that they contain approximately ten times as much explosive energy as all other tactical types (1 and 2) combined.

Given their great power (about one megatonne), their relatively poor accuracy (probably a few kilometers) their peculiar range (covering all Europe and not much more from European Russia) and their vulnerability (they are old, and therefore more vulnerable than modern types to a pre-emptive attack), they would be most effectively used in the early strikes against governmental and major military command centers, major communications centers, large transportation centers (harbors and railyards), manufacturing facilities, and all airports, both civil and military. The history of modern warfare has made all such targets legitimate. Most such targets are in or near large cities, and, in fact, a list of such targets would be, for all practical purposes, simply a list of nearly all the great cities of Western Europe and a lot of smaller ones besides. There are, of course, some targets which are both soft and not very close to a major city, but the number of these is small compared to 600, and therefore such targets cannot 'draw fire' away from the more lucrative and numerous urban targets. In summary, the most reasonable estimate is that the majority of the 600 medium-range, one-megatonne missiles are at this moment targeted against the centers and suburbs of the cities of Western Europe, and that they are likely to be employed in the early phases of a nuclear war.

The probable result of a nuclear war

In order to understand the probable net results of a nuclear war in Europe, we must first examine the results of a single one-megatonne explosion.

At the time of the explosion, huge amounts of energy are released (for one megatonne a total of 4.6×10^{15} joules) in a variety of forms. Table 2.2 gives the proportions of the most important forms of energy for a bomb burst at an altitude of 1 000 or so meters above the ground.

In brief, the energy release in the form of '*blast*' topples buildings and trees. It kills people mainly by throwing loose objects at them, by throwing them at fixed objects, or by crushing them in collapsing structures. The energy released as *thermal radiation* ignites fires in flammable structures, and in fields and forests. It kills people both as the indirect result of such fires and by causing sufficiently serious burns on exposed parts of their bodies. It can also cause temporary or permanent blindness in people who happened to be looking at the site of the explosion when it occurred. The *prompt nuclear radiation* causes death in people and other mammals who happened to be close to the explosion but who were not killed by one or more of the effects named above. The *delayed nuclear radiation* or 'fallout' as it is commonly called, spreads out to large distances, sometimes killing people even hundreds of kilometers away from the site of the explosion, and causing a number of delayed but still premature deaths in remote parts of the world. The numerical dimensions of these destroying and killing effects are summarized in Table 2.3.

Table 2.2. Form of energy release in a one-megatonne air burst

Type	*per cent*
Blast	50
Thermal Radiation	35
Nuclear Radiation:	
(a) prompt	5
(b) delayed	10

Table 2.3. Typical measurements of killing and destroying effects of a one-megatonne nuclear bomb

Effect	*Range*	*Area*
Overpressure, or blast:		
(a) Crater (everything is turned to vapor within about the same range)	200 meters	12–50 ha
(b) Destruction of apartment houses, with death of most inhabitants	6.0 km	100 km²
(c) Destruction of wooden homes, with death of most inhabitants	9 km	250 km²
Nuclear radiation:		
(d) Death due to prompt radiation	2.5 km	20 km²
(e) Nuclear fallout sufficient to kill most persons in the open	Elongated area reaching out to 100 km	150 000 ha
(f) Nuclear fallout sufficient to kill persons sheltered in basements of houses or apartments	Elongated area reaching out to 25–50 km	10 000–40 000 ha
(g) Numbers of persons who die prematurely from leukemia and other forms of cancer in other parts of the world	Worldwide	very roughly 1 000 persons per megatonne
Other:		
(h) Forest blown over by blast	7–10 km	15 000 ha
(i) Dry forest ignited	10–20 km	30 000–120 000 ha

Source: The Effects of Nuclear Weapons, Samuel Glasstone, Ed. (USAEC, Washington, DC, 1962). Arthur Westing, Windham College, Putney, Vermont 05346, USA, private communications.

An accurate estimate of the number of people killed in such a complex event depends strongly on many factors which vary markedly from case to case (population distribution, type of building, civil defense preparations, amount of and response to warning, if any). Moreover, even when all the important circumstances are known in detail, the theory that is used to translate these facts into an estimate of casualties is very weak. Therefore, we are inclined to turn to experience for an answer. There are, of course, only two experimental cases: Hiroshima and Nagasaki. In these two, the bombs were 13 kt and 21 kt respectively, and the casualties (including delayed ones) numbered approximately 100 000 and 50 000. The bombs we are concerned with here are, roughly, 50 times bigger. According to simple theory, such bombs will cause equivalent physical damage in an area about 13 times as large. Translating from Japan 1945 to Europe 1975 is harder. In some ways European cities are tougher: housing is generally sturdier in Europe today than it was in Japan then. On the other hand, today's cities depend much more critically on complex centralized services, and in this sense they may be more vulnerable. European cities also have much higher population densities than Hiroshima and Nagasaki had. My guess is that a 1-Mt explosion would kill at least one million people, providing only that the population of the target area is substantially bigger than that; otherwise the expected deaths will simply be some major fraction of the total population.

As indicated in Table 2.2, about 15 per cent of the energy of the bomb appears in the form of nuclear radiation. The intensity of the local fallout can easily be sufficient to kill human beings and other mammals. The world-wide fallout from one (or even a thousand) one-megatonne explosions will not lead to the immediate death of anyone. However, nuclear radiation does induce leukemia and other cancers, and causes premature death in other ways. The calculations connecting fallout radiation and such premature deaths are extremely difficult to make accurately, but a very rough estimate of 1 000 premature deaths per megatonne exploded may be used to give a feeling for the magnitude of the problem.

All of the above discussion relates to a single explosion. Multiple explosions reinforce each other in various ways to produce much more than the single sum of their effects. One such synergistic effect is 'social chaos'. In the event of a single, isolated explosion, the people in neighbouring cities can all pitch in and help the striken city to recover. In the case of multiple nuclear explosions, there will be no one to help. The number of deaths, especially in the middle term (*ie*, the next weeks and months) will be much higher per megatonne in the latter case, and the prospects for eventual recovery much less.

Another such cooperative effect involves the depletion of the ozone layer surrounding the earth, and the consequent arrival of large amounts of ultra-violet light at the earth surface. This effect is still not at all well understood, but it could perhaps result in extreme forms of sunburn (leading to skin cancer) and severe eye problems in humans, and might seriously effect vegetation and consequently food supplies.

Europeans ought to know

Putting all this together, it is clear that the bombardment of Western Europe by the 600 medium-range ballistic missiles deployed in the USSR (or any important fraction of them) could easily eliminate virtually the entire urban population by blast alone. In addition, if a large portion of these bombs were exploded on the ground, then a major fraction of the rural population could be killed as well by the resulting nuclear fallout. The use of only one percent of the Soviet strategic forces, if added to the full use of the tactical forces, could *double* the explosive power brought into play. And, in addition to the dead in Europe itself, something in the neighbourhood of a million persons could die prematurely in remote parts of the world.

In summary, today's Western Europeans have chosen to buy current political stability by placing the awful risks described above over their lives and their future. Perhaps their choice was inadvertent; perhaps they did not and even today still do not realize what they have done. In that case, it would seem that they ought to know, and that they ought to reconsider their choice in the light of such knowledge.

*6. 'First use' of nuclear weapons**

Ever since the initial development of the atomic bomb, the first use of nuclear weapons has been the subject of intense debate. Even now, some thirty years later, it is an unresolved question as to whether the US should or should not have dropped the bombs on Hiroshima and Nagasaki.

All during the 1950's, nuclear weapons were brandished by Secretary of State Dulles and others as the counter to Sino-Soviet aggression with larger conventional forces. During this period the implicit threat of first use of nuclear weapons was a fundamental element of our foreign and national security policies.

As the Soviet Union procured an H-bomb capability of its own, the threat of a US nuclear attack became less and less credible. During the Kennedy Administration, defense officials increasingly realized that conventional attacks must be deterred or resisted by conventional forces. Nuclear weapons were primarily regarded as useful to deter a nuclear attack, or to be employed only as a last resort when all conventional means had failed.

Moving away from no-first-use

Meanwhile, the 'firebreak' between conventional and nuclear weapons became stronger as the years passed without their use. Even during the long and difficult years of involvement in Vietnam, the US never seriously considered dipping into its atomic stockpile—an option which was becoming less viable each day. When the US signed Protocol II of the Latin American Nuclear Free Zone Treaty in 1968, it agreed not to use nuclear weapons against contracting parties to that Treaty. Thus, the US cold war policy of maintaining the freedom of action to use nuclear weapons whenever it saw

* Shortened version of a paper by Herbert Scoville, Jr. from *Arms Control Today*, July–August 1975.

fit was gradually becoming eroded by custom and even by formal treaty commitment.

Now, however, there is evidence of a major reversal of this trend. Instead of just being reluctant to commit itself to a no-first-use policy, the US has apparently started on a campaign of atomic threats. President Ford, when queried on 'first use' at his 25 June 1975 news conference, avoided a direct response but admitted that in the past eighteen months US security policy had been changed to provide maximum flexibility for such use of nuclear weapons as our national interest might require.

This brandishing of nuclear weapons peaked on 1 July 1975 when Defense Secretary Schlesinger stated at a press breakfast that 'first use' could conceivably involve strategic forces possibly in a 'selective' strike at the Soviet Union. He attempted to differentiate between a 'selective' and a 'disarming' first strike, the latter not being achievable because of the invulnerability of submarine missile forces. The improved counterforce capability which Schlesinger is seeking through greater accuracy and higher yield MIRV warheads will, however, threaten the Soviet fixed, land-based ICBM force—a capability which his predecessor, Secretary Laird, said the US should never seek. Soviet planners cannot avoid thinking that our military is prepared to initiate a nuclear strike against their ICBM force, apparently what Schlesinger wishes them to believe. Since Schlesinger, in testimony before the Senate Foreign Relations Committee in September 1974, characterized a similar Soviet attack against our Minuteman and bomber forces with several thousand megaton-yield weapons as 'limited', the Russians may not be reassured by Schlesinger's term 'selective strike'. He has stated that they should not worry since they can adopt a 'launch-on-warning' posture. If the Russians follow his advice and launch-on-warning, then his 'selective' strike will only destroy empty silos while their warheads will be killing millions of Americans.

The illusion of containing nuclear war

To allay fears over initiating the use of nuclear weapons, Secretary Schlesinger argues that breaching the 'firebreak' between conventional and nuclear weapons by a selective strike would not run a high risk of escalation to a major nuclear conflict. He believes that logic would override psychological pressures on decision makers for such escalation, which would require a conscious policy choice. He further asks why, since conventional escalation was controlled in Vietnam, similar control could not be exercised in the nuclear case. Such arguments ignore the fact that the time available for making decisions will be much shorter in a nuclear conflict. And even Secretary Schlesinger has admitted that 'Soviet military doctrine does not subscribe to a strategy of graduated nuclear response', so escalation may be inevitable regardless of any US restraint. Finally, since the fate of mankind may depend on the result, can we afford the risk that Schlesinger's judgement might be wrong?

The procurement of new counterforce weapons generates pressures for escalation since both sides will know that unless they preempt a major element of their force could be wiped out. While it might be possible to limit a conflict if nuclear weapons were used only in the battlefield situation, it would seem very unlikely, if not impossible, for it to be controlled once even a few strategic weapons were exploded on the homeland of either the US or the Soviet Union. Even a limited nuclear strike would result in millions of casualties and the pressure to retaliate would be tremendous. A flexible strategic capability only makes it easier to pull the nuclear trigger.

The more America threatens to use nuclear weapons as a response to a conventional provocation, the less the US or its allies will remain prepared to deal with such aggression by non-nuclear means. The forward deployment of tactical nuclear weapons in Europe and Korea can lead the US to be inadvertently embroiled in a nuclear conflict. Presidential permission to use nuclear weapons will be difficult to deny when forward-deployed nuclear forces are in the process of being overrun. Tactical nuclear weapons should be available only as a last resort after careful presidential and allied review of all the consequences. Redeployment of nuclear weapons to rear areas would be a sounder method of raising the nuclear threshold than threatening a 'selective' strategic attack. The use of strategic weapons should be decoupled from tactical situations to reduce the risk of a local conflict escalating to a worldwide holocaust.

Undermining the NPT

Furthermore, this rising crescendo of nuclear threats is undercutting efforts to prevent the proliferation of nuclear weapons. The more nuclear weapons appear to have political or military usefulness, the more non-nuclear weapons countries will become convinced that they too must take advantage of these benefits. The US and other nuclear weapons countries are asking non-nuclear states to forego the option of acquiring nuclear weapons, but they are unwilling to make a commitment not to use such weapons even against these non-nuclear countries. The US should make a declaration that it will not use or threaten to use nuclear weapons on non-nuclear weapons states, who are parties to the Non-Proliferation Treaty (NPT). This does not mean that the US must renounce its 'nuclear umbrella', or that allies of the US would be unprotected against aggression by other nuclear states. Furthermore, to the extent that other nuclear nations make similar pledges, the nuclear threat to these countries is further reduced, providing an added incentive for nations to become parties to the NPT. Unless the spread of nuclear weapons can be halted, they will soon be considered as another conventional weapon. American security will then be greatly decreased since nuclear weapons are the great leveler, and the US will become vulnerable to devastation by small nations or even sub-national groups.

*7. Between 'first use' and 'first strike'**

A first-use doctrine is not new—it has been a basic element of US strategy since World War II. As Schlesinger himself said at a news conference on 1 September 1975, 'we have never indicated that we were prepared to renounce the option of first use'. The USSR has also never renounced a first-use option. In fact, Soviet nuclear strategy stresses nuclear war-fighting capability.

The restatement of the first-use policy is one of a number of recent developments in nuclear doctrine, all of which are related to improving the capabilities for nuclear-war fighting. The first nuclear weapons were made to win World War II. Some groups in the USA and the USSR have never given up the idea that a nuclear war could be fought and won in the traditional way. These groups have sufficient political power in both countries to ensure that nuclear warheads and their delivery systems are developed as rapidly as possible to make them ever more suitable for this end.

Accurately delivered low-yield nuclear warheads, multiple warheads on missiles, very large numbers of strategic delivery vehicles, quick-reaction strategic nuclear forces of high penetrativity, flexible strategic command and control, and strategic cruise missiles are unnecessary and even harmful for a strategy of deterrence but are all highly desirable for effectively fighting a nuclear war. The fact that these weapons are being developed and deployed without significant restraint is firm evidence that each of the two great powers is striving for a first-strike capability.

In reality, neither side could 'win' an all-out nuclear war—either now or in the foreseeable future. Even if all the planned improvements in defensive and offensive strategic nuclear forces are made, a high enough number of nuclear weapons would still survive an initial attack to be able to inflict massive damage on the attacker. We can, however, be sure that the present technological situation will not continue for decades. The vast amount of money being spent on research into antisubmarine warfare, for example, is likely eventually to produce spectacular results. Nuclear ballistic-missile submarines, currently the most invulnerable component of the strategic forces, may then be vulnerable to a mass attack.

The risk of an outbreak of nuclear war is much increased by the current official campaigns to reduce the population's fear of the consequences of a full-scale nuclear war. For this reason official attempts to gain support for the counterforce strategy, the first-use policy and the deployment of new tactical nuclear weapons are deplorable.

---

* From *SIPRI Yearbook*, 1976.

*8. Long-range cruise missiles**

The world is now faced with a new set of technological advances already being exploited in weapon applications, which could drastically alter the conduct of both tactical and strategic warfare. An entirely new class of weapon systems is now emerging as the direct result of recent developments in the micro-miniaturization of electronic components and sensors of electromagnetic radiation. Progress in nuclear warhead technology combined with advances in remotely piloted vehicles (RPVs) and cruise missiles appear to be putting such weapons at the vanguard of a profound change in the means of conventional and nuclear warfare.

Recent advances in large-array microcircuitry (large array microcircuits are electronic circuits fabricated on very thin substrates that contain on a few square centimetres the equivalent of many thousands of electronic logic circuits) have made possible an entirely new set of microminiaturized electronic devices and sensors of electromagnetic radiation. In turn, these techniques make possible the construction of guidance systems for long-range self-guided missiles and unmanned delivery vehicles with ranges that are now independent of guidance considerations, can extend to several thousand miles and are limited at the present time mainly by the efficiency of existing small turbofan engines and the energy content of available jet fuel.

These developments have made possible what is essentially a new method for the accurate delivery of tactical or strategic weapons (conventional or nuclear) over long ranges, since cruise missiles can now be fitted with terminal guidance based on terrain matching and recognition, and remotely piloted vehicles with wide-band jam-proof communication links. Thus, although both cruise missiles and RPVs are 'old' weapons, commonly thought of as inaccurate and useful at best for short-range tactical applications, they are now emerging as a new class of very accurate long-range weapon systems with far-reaching future implications. The same basic electronic advances have also made feasible the deployment of a satellite-based global positioning system that would allow a missile to determine its position while in flight with an accuracy of seven to ten metres anywhere on the surface of the Earth independently of the relative positions of the launching and target points. Such a system which could be ready by 1980, would make possible the delivery of warheads over intercontinental ranges with an accuracy of ten metres.

The US Department of Defense is now funding a project for the development of a long-range cruise missile designed to be launched either from the torpedo tubes of a submarine or from an aircraft; it can be adapted either to tactical or strategic use by a mere modular change of warhead (conventional to nuclear), arming and fusing mechanisms, guidance package, engine and fuel tanks.

---

* From *SIPRI Yearbook*, 1975.

More efficient than a ballistic missile

The missile under development is designed with two guidance systems: an inertial navigation system aided by mid-course updating derived from terrain matching, and a secondary microwave radiometry guidance system. This combination of guidance systems can guide the missile to the target, recognize and land on it. A cruise missile is inherently a much more efficient device than a ballistic missile to carry large payloads since it possesses aerodynamic surfaces; given the ability to guide these vehicles precisely to their target, it can be expected that their intrinsic advantages will prove attractive to the military and that their development will be pursued vigorously, even though weapons are not always developed because of their intrinsic advantages.

From the military point of view the long-range cruise missile is a very attractive new weapon system. New guidance systems which are either self-reliant or dependent on satellite-derived position information can improve accuracies and in theory shrink the CEP to zero; in practice, even at the present state of technology, they can achieve CEPs of under 30 metres. Since high accuracy and large payload make the lethality K (K is a function of the yield and accuracy of a warhead and thus a measure of its destructiveness) of the cruise missile quite high (probably in the K= 1 000 region for missiles armed with a nuclear warhead), the kill probability of this weapon against any target will be virtually 100 per cent. That has great advantages from the viewpoint of the military planners because it reduces drastically the number of men, vehicles and support facilities (the so-called logistics train) necessary to destroy a target. Especially for the United States, wishing to maintain military capabilities around the world at the smallest possible cost and the minimum presence of military personnel abroad, long-range cruise missiles or RPVs offer enormous attraction. These inexpensive, highly accurate weapons will be capable of striking both tactical and strategic targets from seaborne platforms or a few judiciously located bases. In addition, the mobility, multiple basing, ease of handling and guidance, and the low cost of maintaining these weapons in a high state of operational preparedness make them almost completely invulnerable from a surprise attack.

Interservices rivalry and the economic imperative

A very important, yet unspoken factor in the development of SLCMs and ALCMs may be the continuing rivalry over future missions and roles between the air force and the navy, the two technology intensive and conscious services. The air force, for example, points with a degree of satisfaction to the conventional support of tactical ground operations that the B-52 performed during the war in Viet-Nam. Development of the SLCM will provide the navy with a similar capability, and this may be one factor that prompted the decision to develop that weapon.

Yet another factor in this decision was, no doubt, the fact that strategic SLCMs were not prohibited by SALT I.

A second broad class of reasons for the development of the long-range cruise missile system is the mere presence of the technological advances that have made it possible. For many aerospace and electronic industrial firms, to develop new technologies and apply each innovation in as many separate products as possible is a matter of economic imperative.

The great equalizer among nations

The UK, France, Japan, India, Sweden, FR Germany and perhaps Italy and Israel all have the technological infrastructure in electronics, airframes and jet engines to produce long-range cruise missiles. Perhaps not all these countries have the will or the financial capacity to enter the field, but the attractive features of this weapon system and its low initial cost will undoubtedly lure military establishments to pressure their governments to acquire it. And as is the case with most weapons, this emulative process will diminish the security not only of the countries which do not have this weapon, but also the national security of the USA, the USSR and all other countries which do.

The continuing development of this weapon by the USA will further damage the chances to arrest the spread of nuclear weapons, since many near-nuclear states will see in the SLCM not only a threat against their security, but an opportunity as well. The accurate inexpensive cruise missile may prove to be the great equalizer among nations, but it may also prove to the ultimate leveller of their cities.

Cruise missiles do not have the psychological barrier of ballistic missiles. ICBMs and SLBMs have become the symbols of nuclear war and global extermination; their use is therefore constrained by their image, in the minds of political and military leaders, as the vehicles of ultimate folly and catastrophe. Cruise missiles, on the other hand, are thought of in a more casual and benign manner. A decision to use them would appear much less forbidding. This lowering of psychological inhibitions coupled with the fact that the same cruise missile can have a tactical mission with a conventional warhead or a strategic mission with a nuclear warhead could make their potential use seem much less threatening. Thus long-range cruise missiles have the potential to make war more casual and less politically responsible; they greatly increase the possibility of war as well as its quick transmutation from a conventional tactical encounter to a nuclear confrontation.

By far the most serious disadvantage of the long-range cruise missile, and especially the submarine-launched version, is its impact on arms limitation efforts. The obvious pre-requisite of any international agreement limiting one or more weapon systems is that the agreement can be verified. The theorem that governs arms limitations is that an agreement on limiting a weapon is possible only if compliance to it can be verified by non-intrusive means, that is, by observation satellites or remote detection of electromagnetic or nuclear radiation. The SLCM now being developed includes two versions, one strategic with a nuclear warhead and one tactical with a conventional warhead, which are externally identical. In addition, since both weapons will be encapsulated in a canister for firing from a torpedo tube, it will be physically impossible to distinguish between the two versions without literally dismantling the weapon. Since this constitutes highly intrusive inspection, it will be unacceptable to both the US and Soviet military establishments.

A new channel to the arms race

The long-range cruise missile now being developed for the US Navy and Air Force is the forerunner of an entire class of new weapons that opens a new channel to the arms race.

In the more immediate future, the deployment of the SLCM and ALCM will further increase unnecessarily the number of nuclear weapons without enhancing the security of any nation; on the contrary, in the political and military ambience that will be generated by that deployment, the relative ease and low initial cost of its production will almost certainly induce several near-nuclear states to attempt acquisition of long-range cruise missiles. It may prove technically difficult for them to arm these missiles with thermonuclear devices, but it is already well within the technological capacity of several countries to develop fission devices small enough to be carried thousands of kilometres by a cruise missile. Such a proliferation of deliverable nuclear weapons will erode even deeper, perhaps irrevocably, the chances for meaningful arms limitation agreements.

Since development of the SLCM and ALCM appears impossible to arrest, although it could be constructive to do so, we believe that the United States and the Soviet Union could negotiate within the framework of the Vladivostok guidelines a 'no first deployment' agreement to cover all cruise missiles with ranges of above 500 km. An example of an agreement banning the deployment of an already developed weapon could show the way for reaching a proper balance between the responsibility of defence establishments to safeguard the security of their nations, and the immediate need to arrest and reverse the arms race.

*9. The militarization of the deep ocean**

The world seems to be on the verge of a large-scale military development of the deep ocean including the ocean floor. Underwater military systems have, it is true, been in existence for decades. But the conventional, that is, diesel-powered, submarines of World Wars I and II were never independent of the ocean surface. They needed to surface in order to take in oxygen for their batteries as well as for communication purposes.

* From *SIPRI Yearbook*, 1968/69.

The nuclear submarines of the last decade have, on the other hand, been independent of the surface. The do not need oxygen for their propulsion mechanism and they can receive communications by trailing antennae submerged several tens of feet below the surface. Their hulls have so far, however, not been able to withstand the pressures below a depth of 2 000–3 000 feet; their normal operating depth is considerably less than 1 000 feet.

New marine technology is now reducing these and other limitations on operations in the deep ocean. This involves extending the capability of man to work as a diver down to 1 000 feet, which includes the whole continental-shelf area; building deep submergence vehicles that soon can go down to 20 000 feet; using new long-endurance power sources under the sea; development of manned underwater stations and their installation at depths of 6 000 feet; deployment of submarine-detection systems on the ocean floor, etc. But whereas technology is, undoubtedly, the short-term factor that determines the speed and extent of underwater military development, it is not the prime mover. That force must be sought elsewhere, in the strategic competition between the super powers, the United States and the Soviet Union.

Underwater nuclear-missile systems

The case for underwater nuclear-missile systems has often been stated in the negative, as a case against the dominant deterrent, land-based missiles. This can be summarized thus. Land-based missiles can be detected by satellite reconnaissance and other means. At the same time the accuracy of attacking missiles is advancing fast and is coming down to a fraction of a mile measured in CEP. In addition, MIRV systems are being developed with the consequence that there will be a disproportion between the number of attacking warheads and the number of landbased missiles in silos. As a result the latter become increasingly vulnerable. The increases in the efficacy of attack appear in general to be out-running the developments in defence.

These are the arguments presented. How far they are justified is another matter. They imply that there is a significant risk that at some point one of the two super powers might attempt a first strike against the land-based missiles of the other. It can be argued that this is a rather fanciful assumption, and indeed that this process of seeing remote and extremely unlikely threats as real and immediate is one of the strong forces behind the technological arms race.

The ability to hide is thus what makes the modern submarine an ideal strategic deterrent from a military viewpoint. Mobility is a key element in the ability to hide, as is, in certain situations, slow speed: below 5 knots the noise of a submarine will be almost indistinguishable from the background noise.

The invulnerability of a fully deployed undersea deterrent system does not mean that no single vessel can be destroyed by an enemy. It means that the entire system as such is non-targetable at a given time and that, therefore, enough vessels will always survive to strike back.

Other advantages often claimed for the undersea deterrent system include: its invulnerability permits longer reaction time in a crisis situation; it will to some extent move the threat of a nuclear exchange from land to sea; because of the risk of accidents, it is preferable to have nuclear weapons in the sea rather than on land.

The disadvantages of undersea military systems

Even from the point of view of a military planner an undersea deterrent force has some disadvantages. Taking a broader view, one may observe the following disadvantages. Until now the deep ocean and the ocean floor have been practically unmilitarized. The only exceptions seem to be US bottom detection systems outside US coasts and, probably, near passages where Soviet submarines enter the two major ocean basins, and, maybe, in the open Pacific. With the introduction of advanced undersea deterrent forces, operating in the whole ocean volume down to the ocean floor, the demand will come for supplementary military installations for detecting enemy forces and, perhaps, for servicing one's own forces. Indeed, this development has already begun. Because of the 'action-reaction' phenomenon, more and more resources will be spent on undersea warfare in the, perhaps vain, hope of finding countermeasures to Polaris and more advanced deterrent forces. Adding to this is the anticipatory belief, common among military planners, that the adversary will always be able to do what one's own side can do. This may lead to a constant search for countermeasures to the new weapons one is developing oneself—a narcissistic arms race.

10. *Anti-submarine warfare**

Since its inception during World War I, anti-submarine warfare (ASW) has been perceived as a tactical (theatre) naval operation aimed at preventing hostile submarines from successfully attacking one's own surface ships. ASW is a defensive reaction to ever evolving counter-shipping weapon systems. The systems and tactics employed in ASW therefore change as a function of the evolution of submarines and the weapons deployed on them. During World War II, for example, straight-running torpedoes with ranges of a few thousand metres were the main offensive weapons employed by diesel-powered submarines. Consequently World War II ASW systems and tactics were designed to maintain a submarine-free zone of comparable width around a convoy of surface ships. Contemporary counter-shipping submarines employ ordnance of considerably longer range and greater

* From *Tactical and Strategic Antisubmarine Warfare*, SIPRI, 1974.

sophistication: homing torpedoes with ranges of 10–20 km and cruise missiles with ranges of 20–400 km or more. To counter the threat to surface ships created by these new projectiles, ASW technology has had to be modified drastically. A second and even more decisive factor that dictated fundamental changes in the tactics and hardware employed in modern ASW was the introduction of the use of nuclear reactors to propel submarines. This meant improvements in speed, endurance and quietness. It rendered the World War II generation of ASW equipment obsolete and created the need for an entirely new research and development field to support the development of consecutive generations of ASW systems designed to cope with the nuclear submarines.

Perhaps the most momentous development in the field of underwater warfare was the introduction about 15 years ago of the ballistic missile-carrying nuclear-powered submarine (SSBN). Unlike torpedo- or cruise missile-carrying counter-shipping submarines, the ballistic missile submarine has a strategic rather than a tactical role. Its operations are therefore not confined to the vicinity of a convoy or a task force: rather it roams submerged in the millions of cubic kilometres of ocean from where it can be within range of its strategic inland targets. It does not seek to approach, but rather avoids, surface ships, since its main operational requirement is to remain undetected and thereby ensure the availability of its ballistic missiles at any instant.

The deployment of SSBNs, first by the United States, then by the Soviet Union and Britain, and more recently by France, introduced a new set of technical and political problems into undersea warfare. Missile-carrying submarines form the sea-based component of a country's deterrent force. It would thus be highly provocative and certainly strategically destabilizing to attack them (or even to have the capability to attack them), even if 100 per cent success were guaranteed: the response would almost certainly be nuclear retaliation with land-based missiles or long-range aircraft.

The deployment of ABM (anti-ballistic missile) systems to protect urban areas was prohibited by the SALT I agreements, the reason being that such systems impede the ability of ballistic missiles to attack urban areas and hence erode the countervalue role of these missiles. Similarly, and ASW system designed to attack missile-carrying submarines could threaten the second-strike capability of these submarines, and would thus be as undesirable as an urban ABM system: both ABM and ASW systems undermine the credibility of deterrence as a viable strategic posture. The institutionalization of deterrence as the mutual strategic posture of the Soviet Union and the United States (and presumably also of France, Britain and China) appears to proscribe any military operation that could threaten the stability of strategic weapon systems on which the credibility of deterrence is based.

Hunter-killer submarines

By far the most effective weapon against a quiet deep-diving nuclear submarine is another submarine, since it can be equally if not more quiet, can occupy the same portion of the ocean as its quarry (thus avoiding the abbreviation of sonar range caused by the oceanic inversion layers and the ever changing seasonal thermocline), and is large enough to accommodate extensive sonar arrays, their power sources and elaborate processors, and anti-submarine weapons. Thus a submarine is an integrated ASW system that can detect, follow, localize and destroy another submarine. Since cavitation noises decrease with increasing depth, a deep-diving submarine is quieter than a similar craft travelling at the same speed but at a lesser depth (or equally noisy at a higher speed). Reduction of noise levels at high speeds and the ability to dive deeply are primary characteristics of hunter-killer submarines.

Only the United Kingdom, the United States and the Soviet Union possess nuclear hunter-killer submarines (SSNs). Britain has seven such craft with four more under construction. The Soviet Union deploys 28 SSNs of various types (with speeds between 25 and 35 knots submerged) and an additional 40 nuclear submarines carrying countershipping cruise missiles (SSGNs) and therefore capable of a dual role (countershipping and ASW). The Soviet Union apparently plans to continue construction of advanced type SSGNs ('Charlie' and 'Papa' class) during the next few years. The United States has 58 deployed SSNs mostly of the 'Permit' and 'Sturgeon' classes (speeds of 30 and 32 knots, respectively) with six more of the latter under construction. In addition, the USA has started producing a new class of 6 900-ton hunter-killer submarine (the 'Los Angeles' SSN-688 class) which is expected to have a submerged speed of 40 knots and a maximum depth of 300 fathoms (550 metres). The US Navy plans to order 32 such submarines by 1976 and the total number may reach 43. Further in the future, the USA is planning to replace all SSNs other than the 'Los Angeles' class with a new craft, designated as CONFORM, that will combine the technological advances attained with the SSBN-688 class with those of the experimental 'quiet' *Libscomb* SSN-685 submarine now under construction. If this plan is achieved, the US Navy may have 68 CONFORM submarines by the early 1980s, since their declared goal is 100 SSNs by that date. France and the People's Republic of China do not have SSNs at this time.

Tactical and anti-strategic submarine warfare

Tactical ASW operations of most national navies are designed to prevent hostile submarines from using the oceans to the detriment of the nation's security in times of war. These tasks involve patrolling coastlines, protecting surface vessels engaged in naval operations and protecting merchant ships performing supply operations. In peacetime, ASW operations are confined to training, monitoring, testing and intelligence gathering. A distinction is necessary between those conventional operations and the ASW tasks facing the US and Soviet navies (and to a lesser extent the British Navy as well). In the case of these powers one must distinguish not only between peacetime and wartime operations, but also between the class of tasks that ASW forces will face in a conventional protracted war and in a nuclear conflict.

Two distinct types of operation can be employed in tactical ASW. One involves the effort to deny the opponent submarines access to large portions of the ocean. This is an offensive operation that attempts to secure entire areas against submarine counter-shipping activities by attacking all submarines that try to enter a given ocean basin (area defence). Patrols by land-based fixed-wing aircraft and helicopters and by roving hunter-killer submarines often lying in ambush outside nuclear submarine bases, and ocean surveillance by fixed arrays are designed to support the operational requirements for area defence. The other type of operation aims at defending a particular point in the ocean occupied by a convoy, a task force or even a single ship such as a carrier, by organizing carrier-based ASW fixed-wing aircraft, ASW helicopters, surface ASW vessels and hunter-killer submarines into protective screens (point defence).

Although detection, localization and classification of missile-carrying submarines can be achieved with the same acoustic systems and the same platforms used in tactical operations, the tactics and purpose of these submarines are distinctly different from those of counter-shipping submarines and ASW operations against them therefore have entirely different scales and goals. First of all, a missile-carrying submarine avoids surface traffic; point-defence ASW is thus irrelevant in this case. Even more importantly, while the purpose of detecting and localizing an attack submarine is to destroy it, the ultimate purpose of anti-strategic submarine warfare (ASSW) is not always the destruction of a missile-carrying submarine. More often than not the last stage of ASSW could be protracted trailing of the quarry rather than the launching of a weapon against it.

The only method that could fulfil the operational requirements for a first strike is the uninterrupted trailing of an opponent's missile-carrying submarines. No other method that can be supported either by existing ASW systems or by any foreseeable technical development can achieve the high confidence required for a first strike. The trailing of nuclear submarines can be practised either by hunter-killer submarines such as those already deployed by the USA, the Soviet Union and Britain, or by special surface or subsurface platforms, designed specifically for the purpose of trailing. Most probably more than one trailer is necessary per missile-carrying submarine if the trail is to be securely maintained.

According to unofficial estimates, the United States could, under favourable conditions, detect and localize most of the Soviet missile-carrying submarines on station or in transit most of the time, using presently available area-defence ASW systems. The deployment of the Moored Sonobuoy System (MSS) and the very large SAS surveillance array will enhance this capability. On the other hand, the Soviet Union possesses enough hunter-killer submarines to be able to trail a substantial number of the Polaris craft during peacetime. As the Soviet naval planners re-assess the relative danger resident in US missile submarines as compared with aircraft carriers, it is probable that they will assign a larger and larger portion of their hunter-killer submarines to the task of trailing the Polaris. Thus the capability for either the Soviet Union or the United States to launch a successful damage-limiting attack against each other's or any other nation's sea-based deterrent is certain.

Systems developed and deployed explicitly for use in area-defence ASW operations offer at the present time the option of a destabilizing damage-limiting attack on either the Soviet or the US submarine missile forces. This threat can be eliminated by the introduction of submarine missiles with rangers greater than 7 000 km, that will allow SLBMs of both great powers to loiter in their respective coastal waters while well within range of their targets.

*11. Nuclear-weapon proliferation**

Few people have a clear idea of how extensive the spread of nuclear technology around the world has already become or how rapidly it will most likely continue. At the end of 1975, 168 nuclear power reactors (generating capacity greater than 20 million watts of electricity (MWe)) were producing a total of about 73 000 MWe in 19 countries (Argentina, Belgium, Bulgaria, Canada, Czechoslovakia, France, FR Germany, German DR, India, Italy, Japan, the Netherlands, Pakistan, Spain, Sweden, Switzerland, the UK, the USA and the USSR). All of these countries except the Netherlands and Pakistan have additional commercial power reactors under construction. China has constructed one or two power reactors. An additional nine countries have their first commercial power reactors under construction (Austria, Brazil, Finland, Hungary, Iran, Mexico, South Korea, Taiwan and Yugoslavia). And many other countries, including Bangladesh, Cuba, Egypt, Indonesia, Iraq, Israel, Kuwait, Libya, Luxembourg, the Phillippines, Poland, Romania, Saudi Arabia, South Africa, Thailand and Turkey, have announced plans to acquire power reactors.

By 1980, if the present forecast is realized, 29 countries will have installed nuclear power reactors with a total electrical generating capacity of about 219 300 MWe, about eleven times the 1970 figure. Looking further ahead, it is probable according to the latest predictions, that the 1980 figure will be multiplied more than sixteen-fold by the year 2000.

Nuclear proliferation and world security

As an inevitable by-product of this nuclear power production, huge quantities of plutonium will be produced each year. Plutonium can also be produced in research reactors and special plutonium production reactors. About 50 countries already operate research reactors. Plutonium can be used as the fissionable material for the production of nuclear weapons. It is this

* From *SIPRI Yearbook*, 1975 and 1976.

link between peaceful nuclear technology and nuclear weapon manufacture which is the key issue in the proliferation problem. In terms of world security, the danger of the proliferation of nuclear weapons is undoubtedly the most disturbing aspect of the spread of peaceful nuclear technology.

The world's nuclear reactors are already producing thousands of kilograms of plutonium each year (the 1975 figure is, in fact, about 25 000 kilograms per year). The rate of production will continue to rise exponentially so that by 1978 they will be producing over 50 000 kilograms annually and by 1982 about 160 000. By 1980, the world will have accumulated about 350 000 kilograms of plutonium. Because ten kilograms of plutonium are more than enough to manufacture one nuclear weapon of 'nominal' (20-kiloton) size, and because plutonium has an extremely high monetary value (higher than gold) and is an exceedingly toxic material, the need for nations to safeguard the plutonium they produce is obvious.

Much fissionable material and nuclear equipment for peaceful purposes is imported and the exporting state usually insists on special assurances that there will be no diversion of the fissionable material produced by the imported nuclear equipment, to military uses. The task of preventing this source of diversion has become the main activity of the International Atomic Energy Agency.

But not even the best safeguards system can ensure that there is absolutely no diversion from peaceful to military activities. At the present rate of nuclear power reactor construction, we are facing a 'bomb-a-week' rate of diversion possibilities even with the best of safeguards technology available. It is therefore necessary to have a credible political barrier, in addition to a technical one, if the prevention of nuclear weapon proliferation is to be fully effective. This was to have been the function of the Treaty on the Non-Proliferation of Nuclear Weapons (NPT) which entered into force on 5 March 1970.

No effective safeguards without nuclear disarmament

The NPT has added a new dimension to nuclear safeguards. Each non-nuclear-weapon state party to the treaty has undertaken to accept safeguards applied by the IAEA to all fissionable material in all peaceful nuclear activities within its territory, under its jurisdiction, or carried out under its control anywhere. In other words, safeguards under the NPT apply not only to imported fissionable material and that produced or processed in imported nuclear equipment but also to other fissionable material produced indigenously. This is a crucial difference between safeguards under the NPT and safeguards outside the NPT.

The Indian nuclear explosion has dramatically demonstrated the present fragility of the NPT as a proliferation preventative. The treaty is weak because two nuclear-weapon powers (China and France) and many important states with ambitious nuclear plans (among others, India, Israel, Brazil, Argentina, Pakistan and South Africa) have not associated themselves with it.

Many other states (such as Egypt and Turkey) have signed but have not yet ratified the treaty. But most serious of all, the nuclear-weapon parties to the treaty (the UK, the USA and the USSR) have failed to fulfil their main obligation under the treaty (Article VI) to take effective measures towards nuclear disarmament.

12. *Safeguards against nuclear proliferation**

Nuclear energy may be a blessing. It can be used to generate electricity; it can serve as a research tool; it can be employed in medicine, in agriculture and in industry, for such various purposes as fighting cancer, improving the yield of cereal plants, combatting insect pests and controlling the quality of machinery. It can also be used in the most devastating weapon ever devised.

For as long as there has been international exchange of information on atomic technology and one state has provided another with nuclear material and equipment there have been demands for some kind of assurance that such supplies would be used as agreed. The most important and obvious aspect of this assurance is contained in the concept of international nuclear safeguards.

The objective of safeguards as administered by the International Atomic Energy Agency is the timely detection of diversion of significant quantities of nuclear material from peaceful nuclear activities to the manufacture of nuclear weapons or of other nuclear explosive devices or for purposes unknown, and deterrence of such diversion by the risk of early detection.

Safeguards have by now successfully outgrown the experimental stage. If safeguards help to delay and prevent proliferation of nuclear weapons, or at least contribute to the confidence that a given state is not using its peaceful nuclear activities for military purposes, they will have justified their existence. They also constitute a unique exercise in international verification, which may eventually set a useful precedent for 'real' disarmament measures, if not applicable in their entirety then in part: as an example, perhaps, of an international arrangement to gather data on the production of certain materials and devices, and of the manner in which use can be made of national systems of accounting and control.

It is sometimes argued that once new techniques, such as laser beams, become practical in the detonation of hydrogen weapons, safeguards against the diversion of nuclear material will have outlived their usefulness. This is like saying that since it is possible to purchase howitzers there is no further need to licence and control handguns. There will be an ever increasing amount of special fissionable materials in circulation, and these will have to be controlled nationally and internationally. The fact that there may be other

* From *Safeguards Against Nuclear Proliferation*, SIPRI, 1975.

more efficient means of mass destruction does not make fission bombs any less dangerous.

Over the longer term there are several potential factors which may make it easier for safeguards to keep up with the increased use of nuclear materials: (1) as more countries acquire complete nuclear fuel cycles, the emphasis on strategically important stages in the cycle will mean a relative simplification of the safeguards effort; (2) as more governments establish sophisticated systems of accounting for and control of nuclear material, the safeguarding organizations will be able to make better use of the information provided; and (3), as experience grows and development leads to better results, improved techniques and more effective instruments can be applied. There may be some validity in the objection that, with the growth of the nuclear industry, a 'confidence level' of 95 per cent or more may still leave a considerable gap and large quantities of material not accounted for. But there is reason to expect that this gap will narrow with improved measurement techniques and more effective measures for physical security. In any case, there would seem to be no alternative. One may reason that *no safeguards* would be preferable to *inadequate safeguards*, because the latter might create a false impression of security. This is true. But, a control that yields a confidence level of somewhat less than 100 per cent is better than no control, as long as one remains aware of the shortcomings of the system.

Will the growing nuclear industry continue to tolerate the application of safeguards? In practice, the 'burden' is much less heavy than originally feared. Safeguards are in any case a minor price to pay, given the inherent dangers of the processes and the materials used. National governments will continue to keep an extremely careful eye on the nuclear industry, against which the additional intrusion by an international body could be easily discounted.

An essential conclusion to be drawn from the experience gained so far with Agency safeguards using the term 'safeguards' to cover a range of activities, from the negotiation of the underlying agreement to the verification of a physical inventory in a facility at the end of a material balance period is that the effectiveness of the entire exercise depends on the cooperation of the state concerned and each of the authorities involved: from the civil servant negotiating the agreement to the facility operator who has to make his inventory accessible to IAEA inspectors. This cooperation will continue to be needed wherever and as long as safeguards will be applied. The Agency has to earn it constantly, by employing a professionally competent approach, tact, discretion and perhaps most importantly by never asking for more information than it demonstrably needs. *Safeguards cannot be imposed.* They can only be made possible by continuing mutual agreement at every level.

13. *The case of peaceful nuclear explosions**

The long and woeful tale of test ban negotiations is not nearing any happy end. What has lately served as a stumbling block in the negotiations machinery is what goes under the name peaceful nuclear explosive devices (PNE, sometimes PNED).

The pretention that some nuclear explosions could be permissible if they envisage only 'peaceful purposes' makes the whole question of PNEs the connecting link between the proliferation and the test ban issues, or to be more specific, between the Non-Proliferation Treaty with its Article V and a Test Ban Treaty, whether Partial as now or Comprehensive as is the goal set by the international community. Article V of the NPT promised that the benefits from any peaceful applications of nuclear explosions would be made available to the non-nuclear parties to the Treaty at low cost. It also stated that:

> Non-nuclear-weapon States party to the Treaty shall be able to obtain such benefits, pursuant to *a special international agreement or agreements*, through *an international* body with adequate representation of non-nuclear-weapon States. Negotiations on this subject *shall commence* as soon as possible after the Treaty enters into force. (Italics added.)

The Treaty did enter into force in 1970. No negotiations have taken place. No special agreement or agreements have been concluded. No international body has been appointed. The negligence to solve these outstanding problems has set spokes in the wheels for all attempts to discipline nuclear developments and their spread. The Indian nuclear explosion of 18 May 1974, said to envisage 'peaceful purposes', is a telling case in point.

Peaceful and non-peaceful nuclear explosions

Two preliminary questions have to be answered: What is the difference between peaceful and non-peaceful nuclear explosions? And what would be the true value of keeping the right to conduct 'peaceful' explosions, thus risking that a comprehensive ban on nuclear testing may have to be sacrificed?

The truth to be kept firmly fixed in our minds is that there is no essential distinction possible between nuclear explosive devices, to be used for military or for civilian purposes, one for bombing some place on earth and one for engineering work to mine or excavate it. All nuclear devices are potential bombs, and of a destructive force way beyond conventional explosives. The sole differentiation that can be claimed is the ephemeral one of intent.

In practice, there will of course appear circumstantial differences at the later steps of disposing of the devices, e.g. in regard to the degree of weaponization, the different detonating procedures selected, the different types of casing for careful emplacement in engineering projects or for firing

* Excerpts from Alva Myrdal: *The Game of Disarmament*, Stockholm, 1976.

by military delivery vehicles, and also the nature of the custodians, whether civilian or military authorities. But this specification appears rather late in the development process. In the early phase of R&D work they are certainly indistinguishable. Typically, the cost of such development of nuclear devices has usually been carried on the military budgets.

The crucial step on the ladder towards manufacture of nuclear weapons is testing. Thus the problem for an agenda on PNEs today must be formulated in terms of both the Non-Proliferation and the Test Ban Treaties, the existing Partial one (PTB) as well as the prospective Comprehensive one (CTB).

The problem must also be formulated as an issue of discrimination or non-discrimination as between the nuclear weapon powers and others. In the NPT, promises were held out that the 'benefits' of peaceful nuclear explosives should be made available by the possessors of this advanced technology to non-nuclear weapon parties to the Treaty presumably on an egalitarian basis with them.

What is then the true value of these 'benefits'? A fair appraisal according to expert opinion may be borrowed from the Soviet Academician V. S. Emelyanov:

> The possibilities of using nuclear explosions for civil purposes have been studied mainly in the United States and the Soviet Union. Both countries have been examining the feasibility for opening up ore fields, for building water reservoirs in arid regions, for earth-moving operations in canal construction and so on. In the United States the Plowshare Program was established to implement a number of such projects; the Soviet counterpart is 'The Programme of use of commercial underground nuclear explosions'. Such studies have so far been largely theoretical, and although much useful data has been obtained from test explosions, none of the projects under investigations has yet reached the stage of wide and practical application. The advantages of using nuclear explosions for such projects lie chiefly in saving labour and therefore money. However, the danger of subsequent radioactive contamination of the environment is very real; the problem of designing a 'clean' explosive has still not been solved. It is concluded that, at present, peaceful nuclear explosions are advisable only for exceptionally urgent problems which cannot otherwise be solved.

What Emelyanov did not touch upon in this context is an account of the tug-of-war that has been going on between the haves and the have-nots on this matter of PNE. This hinges on the question of discrimination. The attitudes of the have-nots oscillate with the evaluations of potential benefits. Either we believe as told in early period of enthusiasm that PNEs might yield tremendous economic advantages—then the lesser powers should be given an equal share in these techniques for enhanced development. Or we assume that the present mode of pessimism is more justified—then the nuclear weapon powers should be equally prepared to renounce PNE, not least in order to close the safety-valve on clandestine weapons testing.

The most dangerous loopholes

As both superpowers have come to agree with experts that the material, economic value of using nuclear explosions for civilian engineering purposes is still quite uncertain, there must be some other reason for not giving them up. This might at least have been done through a temporary moratorium while the subject is further studied. The suspicion now appears unavoidable that when the superpowers have tried to becloud the issue and refused even to enter into the stipulated negotiations on special agreements for their regulation, it must have been in order to shield the right to conduct 'peaceful' nuclear explosions so as to have testing capabilities secured from limitations which are or might be placed on testing of nuclear weapons. They are wanted as a loophole.

This political intention has become manifest when the Moscow 1974 Threshold agreement between the superpowers make an unconditional exception for PNE, protecting them from any limiting whatsoever, even from the limit of 150 kilotons yield for similar testing of nuclear weapons.

The two superpowers indulge in double-talk, confessing on the one hand that 'it is not possible to distinguish between the technology of nuclear devices for peaceful engineering purposes and that for nuclear weapons', but on the other hand that 'it is important to develop a set of agreed procedures for peaceful explosions, as and if they become technically and economically feasible, to distinguish them clearly from weapons tests and make sure that they would not be used for weapons purposes'.

Evidently, hope was held out to satisfy the military and pacify the public —that the two governments might work out 'agreed procedures'—note: not multilateral—for PNE. What would it mean for the international situation and multilateral negotiations if they tried to win sanctioning of this bleak prospect in Geneva or the United Nations? Is it to serve as any model for India and others to follow? This is a most serious question.

14. *Non-governmental nuclear weapon proliferation**

Nuclear weapons are relatively easy to make assuming the requisite nuclear materials are available. Fission explosives can be made with a few kilogrammes of plutonium, high-enriched uranium or uranium-233. The design and fabrication of a simple, transportable, fission explosive device is not a difficult task technically. A variety of radiation weapons are also conceivable, and they are simpler to make than fission explosives. Moreover, what appears to be one of the most effective types of radiation weapon, a plutonium dispersal device, requires only a few grammes of plutonium to make. The effects of the use of a nuclear weapon depend on the characteristics of the device and of the target area. The effects of a relatively simple,

* From *Nuclear Proliferation Problems*, SIPRI, 1974.

low-yield fission explosive or a plutonium dispersal device involving hundreds of grammes can be sufficiently intense and widespread to kill tens of thousands of people and cause hundreds of millions of dollars in property damage.

The use of nuclear energy to generate electric power will result in very large flows of nuclear weapon material through various fuel cycles. By 1980 several thousand kilogrammes of nuclear weapon material will be present in civilian fuel cycles in a number of countries with large nuclear power industries. Thereafter, the amount of nuclear weapon material involved in nuclear power industries throughout the world is expected to increase rapidly for many years.

The analysis of non-governmental diversion possibilities which follows is mainly intended to provide readers with a more informed basis for making their own judgements concerning the credibility of the risks involved—judgements necessarily based on their own views of human nature.

Diversion by one person acting alone

The possible reasons for one person to steal nuclear weapon material from the nuclear power industry comprise a broad spectrum. On the rational end of this spectrum is financial profit and on the irrational end is a psychotic expression of extreme alienation from society as a whole. In between lie such motives as settling a grudge against the management of a nuclear plant or a strong conviction that nuclear weapon proliferation is a good thing. Profit appears to be by far the most sensible general motive for an individual to steal nuclear material.

Diversion by a profit oriented criminal group

There are two reasons why a profit oriented criminal group might want nuclear weapon material. One is obvious: the realization of large profits in black market or ransom dealings. The other reason is less obvious: the possession of a few nuclear explosives or radiation weapons might place a criminal group effectively beyond the reach of law enforcement authorities. A criminal organization might use the threat of nuclear violence against an urban population to deter police activities directed specifically against nuclear theft. A criminal group might also use nuclear weapons to coerce from the police a tacit or explicit grant of immunity from law enforcement for a broad range of other lucrative criminal operations.

Diversion by a terrorist group

While financial gain should not be excluded entirely, the dominant motive of a terrorist group which might attempt to divert nuclear material would probably be to enhance its capabilities to use or threaten violence. An important, through secondary, purpose might well be to provide itself with an effective deterrent. In these respects, a terrorist group which possessed a few nuclear explosive devices would be in a qualitatively different position offensively and defensively than one with only conventional arms. Diversion could place nuclear weapons in the hands of groups that are quite willing to resort to unlimited violence.

Conclusions

Non-governmental nuclear weapon proliferation could adversely affect the security of governments and people generally and the prospects for the future development of nuclear power. Nuclear-weapon and non-nuclear-weapon states have a common interest in the effectiveness of national safeguards systems to prevent this type of nuclear proliferation. Thus, close international coordination and cooperation is needed urgently to develop and implement national systems. In the design of national safeguards, the use of the best available technology and institutional mechanisms is justified in view of the dangers of non-governmental nuclear weapon proliferation.

15. *Nuclear-weapon accidents**

Accidents involving nuclear weapons are important for two reasons:

(a) They might start a nuclear war. This could happen if one country detonated a bomb by accident on the territory of a nuclear power or a nuclear power's ally. It might also happen if it dropped a bomb on its own territory, and another country was suspected.

(b) An accidental detonation, even if it did not start a nuclear war, could do great damage if it were detonated over a populated area.

Accidents could result from some kind of mechanical failure, or from the miscalculation or insubordinate behaviour of members of the military forces who operate the weapons delivery systems.

The one case of insubordinate behaviour that could be found in the literature is, in a sense, an example in reverse. The fourth French nuclear weapon test was in preparation when the 'Revolt of the Generals' of the French military forces in Algeria took place. The French scientists at the test site were apparently fearful for the security of their incipient nuclear explosive. They began hurried preparations to detonate the device, which they did against the wishes of the military commander of the test site, to forestall any possibility of its capture by dissident military forces.

* From *SIPRI Yearbook*, 1968/69.

The possibility of insubordinate behaviour is treated sufficiently seriously in the United States for there to be a standing research programme about it. In each year from 1963 to 1966 the annual report of the US Atomic Energy Commission had the phrase: 'Research was also conducted with the objective of providing improved devices for installation in nuclear weapons to prevent unauthorized employment'.

The Goldsboro accident

Perhaps the single most important example in the published literature of an accident which nearly resulted in a catastrophe occurred in 1961 at Goldsboro in North Carolina. Dr Ralph Lapp, who had been head of the nuclear physics branch of the Office of Naval Research, wrote:

> According to a study of the accident problem made by an independent, non-military group, nuclear weapons have been involved in about a dozen major incidents or accidents, mostly plane crashes, both in the United States and overseas. In one of these incidents, a B-52 bomber had to jettison a 24-megaton bomb over North Carolina. The bomb fell in a field without exploding. The Defense Department has adopted complex devices and strict rules to prevent the accidental arming or firing of nuclear weapons. In this case the 24-megaton warhead was equipped with six interlocking safety mechanisms, all of which had to be triggered in sequence to explode the bomb. When Air Force experts rushed to the North Carolina farm to examine the weapon after the accident, they found that five of the six interlocks had been set off by the fall! Only a single switch prevented the 24-megaton bomb from detonating and spreading fire and destruction over a wide area.

In a more recent volume Dr Lapp reiterates that the B-52 in the Goldsboro accident carried two 24-megaton bombs, and that as a result of the accident President Kennedy initiated a review of procedures and mechanisms for weapons safety.

Numbers of accidents

There are various estimates of the number of accidents which have involved nuclear weapons. Immediately following the 23 January, 1968 B-52 accident in Greenland, the US Department of Defense issued a list of 'previous accidents involving nuclear weapons carried on Air Force planes'. It listed twelve accidents from February 1958 to January 1966. The item discussed by Dr Lapp is listed as 'January 24, 1961. A B-52 from Seymour Johnson Air Force Base, Goldsboro, North Carolina, carrying unarmed bombs crashed 15 miles north of the base.' The meaning of the term 'unarmed', which is referred to in most of the other accidents too, is uncertain. It appears to refer to the condition of the bombs before, rather than after, the accident.

In addition to the accidents in the list released on this occasion by the Defense Department, three other accidents had been publicly reported on previous occasions. According to Dr Lapp, there had also been acknowledged accidents overseas at:

—a north African base (Morocco),
—in England,
—off the US Atlantic coastline,
—in the Arctic.

There have also been an unspecified number of accidents when intercontinental ballistic missiles, presumably fitted with nuclear warheads, have been destroyed by fire or explosion. One operational ICBM blew up on its launching pad. Anti-aircraft missiles have misfired several times and have been accidentally launched at least twice.

Altogether these and other sources furnish a total of at least 33 major accidents up to March 1968. However, there are sources which suggest a higher number. One source refers to 'lesser accidents', involved in the maintenance, transportation, or modernisation of actual nuclear weapons which are known to have occurred, and it places the number of these at about fifty for US weapons since WWII. Another total has recently been given in relation to the 1961 investigation ordered by President Kennedy after the Goldsboro incident. President Kennedy was then reportedly told that since the end of WWII there had been more than 60 accidents involving nuclear weapons—including two cases in which nuclear-tipped anti-aircraft missiles were inadvertently launched.

There are reasons for thinking that the total number of accidents involving nuclear weapons systems is significantly higher than the number officially announced.

It is noticeable that all recorded nuclear-weapon aircraft accidents have been of long-range bombers. No accidents have been recorded involving nuclear weapons and carrier or land-based fighter bombers, ASW aircraft, or for other smaller tactical weapons of naval and land forces. It is true that such weapons are likely to be loaded with nuclear weapons or put on alert rarely, compared with United States long-range bombers which are known to have been flown loaded, on airborne alert, in large numbers. But zero accidents involving nuclear weapons would be a remarkable record.

Soviet accidents

The great majority of the accidents listed are of United States weapons systems, and there is little doubt that some unknown quantity of similar accidents from other nations is missing, presumably from the USSR as the second major nuclear power. However, information is no better concerning the three remaining nuclear powers. The reason that we have US examples is that they are either announced by the US Government or revealed by other sources of the Western press. On the other hand 'The Soviet Union, Great Britain, France and the People's Republic of China never mention publicly their own safety programs or anti-accident techniques . . . [or]

mechanical or personal mishaps'. At the same time one must anticipate fewer accidents from nations whose nuclear arsenal is not kept in as high a state of readiness as that of the United States.

Only two reports of Soviet accidents have been found. 'Premier Kruschev is reliably reported to have told Vice-President Nixon about an erratic Soviet missile which was destroyed by a signal from the ground as it headed towards Alaska'. The source does not indicate if this was an operational ICBM or a test missile of another sort. Another reference, for which no corroboration has been found, states:

> 'Helsinki—A tremendous explosion is reported to have blasted a Soviet ICBM base near Alakkrtti close to the Russo-Finnish border. Sources said they believed the blast was a nuclear explosion'.
> (*Missiles and Rockets*, Vol. 6, no. 10, 7 March 1960).

3. Chemical and bacteriological warfare

1. *UN report: chemical and bacteriological (biological) weapons and the effect of their possible use**

On 1 July 1969 the UN Secretary-General published a report prepared by a fourteen-member group of experts on 'Chemical and Bacteriological (Biological) Weapons and the Effects of their Possible Use'. Following is a brief summary of the report.

No form of warfare has been more condemned than has the use of CB weapons.

Most of the knowledge concerning the use of chemical weapons is based upon the experience of the First World War. The agents used in that war were much less toxic than those which could be used today, and they were dispersed by means of relatively primitive equipment as compared with what is now available and in accordance with battlefield concepts of a relatively unsophisticated kind.

Since the Second World War, bacteriological (biological) weapons have also become an increasing possibility. But because there is no clear evidence that these agents have ever been used as modern military weapons, discussions of their characteristics and potential threat have to draw heavily upon experimental field and laboratory data and on studies of naturally occurring outbreaks and epidemics of infectious disease, rather than on direct battlefield experience.

The outstanding characteristic of these weapons, particularly the biological, is the variability—amounting under some circumstances to unpredictability—of their effects. Depending on environmental and meteorological conditions, and on the agent used, the effects might be devastating or negligible. They could be localized or widespread. They might bear not only on those attacked but also on the side that initiated their use, whether or not the attacked military forces retaliated in kind. Civilians would be even more vulnerable than the military.

To appreciate the risks of biological warfare, one has only to remember how a natural epidemic may persist unpredictably and spread far beyond the initial area of incidence, even when the most up-to-date medical resources

* From *The Problem of Chemical and Biological Warfare*, Vol. IV, SIPRI, 1971.

are used to suppress the outbreak. The difficulties would be considerably increased were deliberate efforts made to propagate pathogenic organisms.

Chemical and biological warfare agents

Chemical agents of warfare are taken to be chemical substances—whether gaseous, liquid, or solid—which might be employed because of their direct toxic effects on man, animals and plants. Bacteriological agents of warfare are living organisms, whatever their nature, or infective material derived from them, which are intended to cause disease or death in man, animals, or plants, and which depend for their effects on their ability to multiply in the person, animal, or plant attacked.

All biological processes depend upon chemical or physico-chemical reactions, and what may be regarded today as a biological agent could, tomorrow, as knowledge advances, be treated as chemical. Because they themselves do not multiply, toxins, which are produced by living organisms, are treated in the report as chemical substances. As a class, chemical agents produce their injurious effects more rapidly than do bacteriological agents. The time between exposure and significant effect may be minutes, or even seconds, for highly toxic gases or irritating vapours. Blister agents take a few hours to produce injury. Most chemicals used against crops elicit no noticeable effect until a few days have elapsed. On the other hand, a bacteriological agent must multiply in the body of the victim before disease (or injury) supervenes. This period is rarely as short as one or two days, and may be as long as a few weeks or even longer.

The effects of most chemical agents that do not kill quickly do not last long, except in the case of some agents such as phosgene and mustard, where they might continue for some weeks, months, or longer. On the other hand, bacteriological agents that are not quickly lethal cause illness lasting days or even weeks, and on occasion involve periods of prolonged convalescence. The effects of agents which act against plants and trees could last for weeks or months and, depending on the agent and the species of vegetation attacked, could result in death.

Because they infect living organisms, some bacteriological agents can be carried by travellers, migratory birds, or animals to localities far from the area originally attacked. The possibility of this kind of spread does not apply to chemical agents. But control of contamination by persistent chemical agents could be very difficult. Should large quantities of chemical agents penetrate the soil and reach underground waters, or should they contaminate reservoirs, they might spread hundreds of kilometers from the area of attack, affecting people remote from the zone of military operations.

In circumstances that favour their persistence, herbicides, defoliants and perhaps some other chemical agents might linger for months, stunting the growth of surviving or subsequent plant life, and even changing the floral pattern through selection. Following repeated use, certain chemical agents could even influence soil structure.

The experts described some of the properties of chemical agents affecting man and animals: nerve agents; blister agents (vesicants); choking agents; blood agents; toxins; tear and harassing gases; psychochemicals; as well as agents affecting plants—herbicides (defoliants). They also gave the characteristics of the methods of delivery of those agents.

The following calculations were presented of possible effects of a nerve gas attack on a city:

The population density in a modern city may be 5 000 people per square kilometre. A heavy surprise attack with non-volatile nerve gas by bombs exploding on impact in a wholly unprepared town would, especially at rush hours, cause heavy losses. Half of the population might become casualties, half of them fatal, if about 1 ton of agent were disseminated per square kilometre. If such a city were prepared for attack, and if the preparations included a civil defence organization with adequately equipped shelters and protective masks for the population, the losses might be reduced to one half of those which would be anticipated in conditions of total surprise.

As in the case of chemical agents, estimates were given of possible consequences of a biological attack, and the conclusion reached was that large-scale bacteriological attacks could have a serious impact on the entire target country. Depending on the type of agent used, the disease might well spread to neighbouring countries.

Whatever might be done to try to save human beings, nothing significant could be done to protect crops, livestock, fodder, and foodstuffs from a chemical and bacteriological weapons attack. Persistent chemical agents could constitute a particular danger to livestock. Water in open reservoirs could be polluted as a result of deliberate attack, or perhaps accidentally, with chemical or bacteriological weapons. The water supply of large towns could become unusable, and rivers, lakes and streams might be temporarily contaminated.

Comparative estimates were made of disabling effects of hypothetical attacks on totally unprotected populations using a nuclear, chemical or bacteriological (biological) weapon that could be carried by a single strategic bomber. According to those estimates the area affected would be: in the case of a nuclear weapon (1 megaton)—up to 300 sq. km; in the case of a chemical weapon (15 tons of nerve agent)—up to 60 sq. km; in the case of a biological weapon (10 tons)—up to 100 000 sq. km.

Production of weapons

Today a large number of industrialized countries have the potential to produce a variety of chemical agents. Many of the intermediates required in their manufacture, and in some cases even the agents themselves, are widely used in peacetime. Such substances include, for example, phosgene, which some highly developed countries produce at the rate of more than 100 000 tons a year; ethylene-oxide, which is used in the manufacture of mustard gases, is also produced on a large scale in various countries; mustard gas and nitrogen mustard gases can be produced from ethylene-oxide by a relatively simple process. Similar remarks were made with regard to biological agents.

Despite the fact that the development and acquisition of a sophisticated armoury of chemical and bacteriological weapons systems would prove very costly in resources and would be dependent on a sound industrial base and a body of well-trained scientists, any developing country could in fact acquire, in one way or another, a limited capability in this type of warfare—either a rudimentary capability that it developed itself, or a more sophisticated one that it acquired from another country.

Conclusion

All weapons of war are destructive to human life, but chemical and bacteriological weapons stand in a class of their own as armaments that exercise their effects solely on living matter. The idea that bacteriological weapons could deliberately be used to spread disease generates a sense of horror. The fact that certain chemical and bacteriological agents are potentially unconfined in their effects, both in space and time, and that their large-scale use could conceivably have deleterious and irreversible effects on the balance of nature, adds to the sense of insecurity and tension. Considerations such as these set them into a category of their own in relation to the continuing arms race.

The present inquiry has shown that the potential for developing an armoury of chemical and bacteriological weapons has grown considerably in recent years, not only in terms of the number of agents, but also in their toxicity and in the diversity of their effects. At one extreme, chemical agents exist and are being developed for use in the control of civil disorders; others have been developed in order to increase the productivity of agriculture. But even though these substances may be less toxic than most other chemical agents, their ill-considered civil use, or use for military purposes, could turn out to be highly dangerous. At the other extreme, some potential chemical agents that could be used in weapons are among the most lethal poisons known. In certain circumstances the area over which some of them might exercise their effects could be strictly confined geographically. In other conditions some chemical and bacteriological weapons might spread their effects well beyond the target zone. No one could predict how long the effects of certain agents, particularly bacteriological weapons, might endure and spread and what changes they could generate. No system of defence, even for the richest countries of the world, and whatever its cost, could be completely secure.

The momentum of the arms race would clearly decrease if the production of these weapons were effectively and unconditionally banned. Their use, which could cause an enormous loss of human life, has already been condemned and prohibited by international agreements, in particular the Geneva Protocol of 1925, and, more recently, in resolutions of the General Assembly of the United Nations.

2. *Possible applications of CB weapons*[*]

CB weapons have the following advantages. The range of effects attainable with them is adaptable to a wide choice of weapon-delivery systems and, provided the weapons will work as predicted, to a wide variety of military situations. They are area and search weapons that can harm an enemy whether he is widely dispersed or concentrated within fortified positions. They can be used against point targets without accurate location of the target. They do not destroy buildings, communication or transportation facilities, or power sources. They have singular area-denial possibilities. They are suited to clandestine or surprise operations on a large or small scale. Their logistic requirement may be small.

But they have several disadvantages. Their use is illegal and potentially outrageous and may thus have damaging political consequences. Their operation is complicated, requiring special training, special safety precautions and special forms of field intelligence. The weapons are subject to unpredictabilities which may sometimes amount to uncontrollability. Their effects may not be confinable to their immediate target area, either in space or in time. These uncertainties will complicate forward planning.

The assets may be considered to outweigh the liabilities, or vice versa, according to the circumstances of a particular mission or conflict.

Strategic employment of CB weapons

Strategic applications can be envisaged for CB weapons both before and after belligerents declare or openly acknowledge that they are at war. During the former period, whether it is succeeded by overt war or not, CB weapons might be considered to lend themselves to covert strategies of subversion or economic warfare. The insidious effects of many CBW agents, particularly infective ones, make them suited to sabotage, for not only may they cause widespread damage, but their delayed effects may also enable the saboteur to escape detection. This is one of the few contexts in which contagious-disease agents seem to hold out much military attraction, for an attacker might reckon the resemblance between a natural and an unnatural epidemic to be close enough to divert suspicion. Recurrent acts of terrorism and assassination, successive crop failures, outbreaks of disease or food poisoning that overextend public health services, all or any of these may spread alarm and despondency, foster disaffection with a ruling régime, or weaken a country's industrial capacity.

Once overt hostilities had begun, CB weapons might find both covert and overt strategic applications, although here the sanctions of retaliation and reprisals would presumably weigh even heavier than before.

---

[*] From *The Problem of Chemical and Biological Warfare*, Vol. II, SIPRI, 1973.

Tactical employment of CB weapons

We describe only those tactical applications of CB weapons that have received frequent attention from military commentators, or which have historical precedent. Quick-acting antipersonnel chemicals could find a variety of applications as fire support for offensive theatre operations. It is possible to conceive of several tactical situations where their area-effectiveness, search-out capacity, persistency, and even their ability to cause casualties without also causing gross physical destruction, could offer considerable advantages over conventional fire support. In this context, they are perhaps more likely to be used in conjunction with other weapons, for the effectiveness of each might thereby increase. This applies both to casualty agents, such as nerve gas, and to harassing agents. One occasion when the synergetic effects might be especially marked would be if nerve gas were used alongside nuclear weapons: the havoc created by the latter might gravely weaken an enemy's overall antigas posture. Likewise it has been demonstrated on battlefields in the past that harassing agents can substantially increase the efficacy of conventional antipersonnel fire against entrenched, but unmasked, troops.

Where the immediate aim is to neutralise relatively small enemy positions, the high casualty rates obtainable with nerve gas might be considered attractive. For greatest effect, nerve-gas weapons would be used in surprise attacks of short duration but high intensity, so that the enemy was given little opportunity to activate protective equipments or don protective clothing. This was a technique with which the British, with their 4-inch Stokes mortars and Livens Projectors, were particularly associated during World War I. Nowadays it might be performed with tactical rockets or missiles, multiple rocket launchers, ground-support aircraft spray tanks or cluster weapons, or time-on-target artillery fire. In offensive operations, the target area chosen for such attack would presumably lie in the area chosen for penetration, and extend throughout the depth of the battlefield.

Persistent agents might be used against target areas that the attacker or friendly forces did not intend to enter for some time, for example, defended positions to be bypassed, or artillery that was not placed in the direction of attack. This was a technique first developed by the Germans with their World War I mustard-gas weapons. Persistent agents might also be fired onto the ground behind enemy troops to delay their withdrawal, a technique which the Japanese used against the Chinese at Ichang in 1941, employing mustard gas and lewisite.

The Italians sprayed mustard gas from aircraft to protect the flanks of their advancing columns during their invasion of Ethiopia.

Chemical weapons might be used for more limited goals than inflicting high casualty rates or denying terrain. Small numbers of them, fired sporadically, could be used to harass the enemy by exposing him to unexpected gas clouds that drifted over his positions. This would force him into respirators, thus lowering his overall combat efficiency. Either casualty or harassing agents could be used for this purpose, the latter threatening acute discomfort rather than death among unprotected personnel. This was one of the original functions of chemical artillery weapons during World War I. A much-publicized use to which US forces have put the irritant agent CS in Viet-Nam (although by no means the most common one) has been to incapacitate a target population suspected of containing intermingled enemy personnel, and then rapidly moving in before the effects of the CS wear off.

Antiplant and anti-animal CB weapons

The tactical applications of antiplant and anti-animal CB weapons are confined mainly to chemical antiplant agents, although anti-animal agents might offer some attractions against an enemy that relied at all heavily on draught animals (horses, mules, camels, etc.) for theatre transportation, communications or supply.

Chemical antiplant agents have been used extensively in the Viet-Nam War by US and South Viet-Namese forces. Defoliants have been sprayed over natural vegetation, particularly jungle cover along communication routes and over suspected enemy bases. The purposes of this have been to facilitate target acquisition and aerial reconnaissance, and to diminish the risk of ambush. The value of these techniques is limited by the delayed action of the agents, and an adequate assessment of their overall costs and benefits to their users in Indo-China has yet to be published. The same applies to the employment of herbicides to destroy food cultivations, which is the other main use that has been made of antiplant chemicals.

The initiation of CBW is illegal under conventional and customary international law. Nonetheless it is conceivable that a country might still consider maintaining stockpiles of CB weapons, or setting out to acquire them, on the basis of their first-use potentialities. There can be little doubt that the initiation of CBW will nowadays be an act of great political moment. CB weapons are exceptionally repulsive to many people. A country that initiates CBW within such a climate of opinion must anticipate sharp political and diplomatic repercussions. It might be supposed that these repercussions would countervail the military advantages to be gained from CBW.

*3. The claims that CB weapons are less inhumane than other weapons**

On many occasions since World War I—and even before it—the claim has been made that CB weapons provide a means for decreasing the barbarity of war. It has been argued that CB weapons can cause less human suffering

* From *The Problem of Chemical and Biological Warfare*, Vol. V, SIPRI, 1971.

than other types of weapon, and that they therefore provide the more humane alternative. This argument has been used on almost every occasion when the protagonists of CB weapons have tried to advance their cause or have come under attack. It has been used to justify actual use of CB weapons in combat, to promote claims for greater support of peacetime CBW programmes, and to resist drives for CB disarmament.

The first strongly-motivated use of the humanity argument was by wartime propaganda writers in Germany during World War I. Stimulated by the outcry raised by their counterparts in Britain against German use of poison gas in April 1915, they produced a succession of newspaper articles and radio broadcasts which maintained that gas was less horrible than explosives because it did not mutilate its victims and that in contrast to the British claims it produced a rapid and painless death.

The fact that the comparison can be made one way—with nastier weapons—does not prove that CB weapons are generally 'humane'. It is merely a piece of special pleading. And different people might rank their preferences in different orders. Take the irritants, for example: during World War I people exposed to them occasionally had to be restrained from shooting themselves in order to escape from their effects; and others were driven mad by the pain and misery caused by the agents.

Assessment

Humanity in war is inevitably an imprecise and perhaps rather paradoxical concept. Allowing for all that, it can be seen that there are a number of fundamental defects in the humanity arguments that have been put forward for CB weapons:

1. The term humane is applied to weapons, not to the context or outcome of their use, despite the fact that the weapons under consideration—generally irritant or incapacitant weapons—can usually be used with humane or inhumane results, according to what other weapons are used in follow-up, how they are used, the temper of the conflict and so on.
2. The arguments often consist of saying that a CBW agent will be used instead of something worse ignoring the fact that, if they are permitted, they are equally likely to be used instead of something less unpleasant.
3. The arguments often imply that a particular type of agent should be permitted for the sake of its humane application, ignoring not only the points above, but also that if one type of agent is admitted all agents are more likely to be admitted and, secondly, that if you admit today's CBW agents you are admitting those developed in the future, too, whatever they may be.

It may seem harsh to imply that the military should be denied the possibility of using in war irritant agents the effect of which, in a suitably restrained condition, could be considered humane. The same applies to incapacitants in so far as their use in some situations could be regarded as humane. For this reason it has been suggested that some sort of exception should be made to a CB disarmament treaty which permitted the possession of some or all of these so-called 'nonlethal' weapons, and allowed their use under specified conditions. But any such proposition implicitly rests on the assumption that on the whole, taking all armies that are likely to be involved in wars and all soldiers down to the lowest rank to which this kind of decision would have to be delegated, the soldiers in making their decision would pay attention to saving the lives and limbs of their enemies, both military and civilian. Only then could the effect be a saving in life and suffering. Unfortunately, the experience of wars in which irritant agents have been used, notably the Sino-Japanese War and the Viet-Nam War, tells in the opposite direction. It might be objected that, in Viet-Nam and before, an inadequate effort or no effort was made to limit irritant agents to humane usage, so that the humanity argument for the weapons was unnecessarily brought into disrepute. But this is no answer since it is unreal to suppose that the pressures which caused the restraints to break down in these cases—pressures coming into play in the field and right up to the top of the political system—would not come into play in other wars too. The clear lesson seems to be that irritant agents, and therefore incapacitating agents too, should be kept out of places where there is a risk that fighting will become unrestrained —and war is as good a description of those places as one is likely to find.

4. *The non-use of chemical weapons during World War II**

Modern chemical weapons were conceived during World War I, when they were used widely by most of the belligerents. They have become enormously more potent since then, but have never been used on so substantial a scale.

At the outbreak of World War II, few people were prepared to say that chemical weapons would not be used at some point in the coming conflict. In the Far East, Japan was using gas in China, and three years earlier a European power, Italy, had been doing the same in Africa. Both sides credited their opponents with massive stocks of chemical weapons and with accelerated programmes for procuring more. Substantial sectors of the civilian population in Europe had been undergoing anti-gas training courses for the past five years. Yet by the time the war ended the massive arsenals of chemical weapons remained substantially untouched. It is important today to form some idea about why these weapons were not used. Many of the constraints operating during World War II exist now, even if in some cases they have been modified by events during the intervening years.

At the outbreak of World War II, most of the doctrine on the combat functions of chemical weapons was derived from the experience of World

* From *The Problem of Chemical and Biological Warfare*, Vol. I, SIPRI, 1971.

War I, expanded and modified by such lessons as had been drawn from the various occasions when chemical weapons had been used during the inter-war period. As regards land warfare, it was taught that chemical operations could usefully be integrated into offensive planning for the following purposes: (*a*) softening up an enemy position prior to assault; (*b*) neutralizing enemy artillery concentrations; (*c*) using persistent agents, such as mustard gas, to tie down an enemy-held area that was not to be attacked; (*d*) protecting the flanks of an advance with barriers of persistent agent; (*e*) using persistent agents to seal off reserves held in the enemy's rear, or to restrict their movement forward; and (*f*) using persistent agents to block enemy lines of retreat. Likewise chemical weapons could be used on the defence (*a*) to engage enemy troop concentrations massing for an assault; (*b*) to neutralize enemy artillery; and (*c*) to contaminate evacuated terrain. The advantage which gas was thought to possess in these roles as compared with conventional weapons lay, first, in the ability of toxic clouds to endanger rather large areas and to penetrate positions protected against conventional firepower, and, secondly, in the ability of persistent agents spread as ground contamination to threaten casualties long after dissemination. But these advantageous effects could be obtained only if both operational and meterological circumstances were favourable.

One explanation for the undecided and generally lukewarm attitudes adopted by military establishments towards CW is to be found in the inherent technical limitations of chemical weapons. In use, they depended closely on the weather, which meant that munitions requirements could only be calculated in advance within rather wide limits: quite a small change in the weather, or an unusual target topography, could demand a ten or twenty-fold increase in the weight of weapons needed for a given effect. As a corollary of this, the results of a given attack in terms of, say, the number of casualties could also vary between wide limits, so that in this sense the effects of chemical weapons were not closely predictable, particularly if there was uncertainty about the level of enemy anti-gas protection.

All in all, the employment of chemical weapons would very considerably aggravate command and control problems, while the sheer bulk of material called for would greatly inflate the supply services. These considerations undoubtedly provided severe disincentives to any military establishment contemplating the initiation of CW, at any rate against a major power.

Military incentives and psychological constraints

World War II thus ended with an accumulation of at least half a million tons of chemical weapons remaining virtually untouched by the belligerents. But the fundamental reason for non-use of chemical weapons almost certainly lay at a deeper level than military inexpediency.

In the first place, there were undoubtedly some situations in which the military arguments for using chemical weapons would have been strong.

If military expediency were the dominant consideration in CW policy-making, material and troop capabilities would have been built up in readiness for such situations. But, in the immediate pre-war period and during the early part of the war itself, CW preparedness consisted for the most part of elaborate defensive preparations and more or less hastily improvised manufacturing programmes for chemical weapons: what little instruction was given to troops in CW was given grudgingly and unenthusiastically by the military establishments, and was almost exclusively concerned with the defensive aspects of CW. The situation was that senior military personnel were unwilling to see merit in gas as a weapon, but were prepared to believe that potential or actual enemies did—or at least were prepared to concede that use of gas might be sufficiently advantageous to an enemy to demand the preparation of defensive countermeasures, and possibly even some sort of retaliatory capability. The reluctance to accept gas as a useful weapon stemmed at least as much—probably far more—from institutional pressures and psychological constraints as from rational considerations of its military utility. A general propensity to believe the worst about enemies, coupled with faulty intelligence, led to a willingness to believe that the other side favoured gas. As the war got under way the major belligerents all suspected their enemies of a readiness to initiate CW, whereas in fact none of them had any serious intention of doing so. Under the stimulus of these suspicions, the growing stockages of chemical weapons were advertised as retaliatory CW stockpiles in the hopes of constraining enemy initiation.

The idea of retaliatory CW stockpiles as deterrents for enemy initiation gained in strength as each belligerent came to realize that retaliation could well be escalatory: a chemical mortar action in some distant combat theatre, even with irritant-agent projectiles, might be met by the gas-bombing of a capital city. Each belligerent was aware of his, or his allies', vulnerability to counter-civilian gas attack, and each belligerent had over-estimated his enemy's offensive CW capabilities. Even though the belligerents' chemical-weapon stockpiles were in fact probably insufficient to cause significant damage—at least until the last two years of the war—each belligerent was sufficiently deterred through his own perceptions of them to become still more disinclined to consider initiating CW. It was only in the last months of the war that the nation with the largest deterrent stockpile, the USA, began to consider turning it to operational use. But it did so in the awareness that other weapons would probably be more useful, and without sufficient enthusiasm to overcome the constraints that it had created for itself: the logistical complications that had grown out of its theatre commanders' unwillingness to maintain forward-area CW stores, and the no-first-use policy that had been declared earlier by its leaders to enhance the credibility of its CW deterrent.

Yet although fear of retaliation goes a good part of the way in explaining why gas was not used during World War II, it was only one factor among several. As a constraint, it was undoubtedly strong, but in fact there was little for it to constrain. With the possible exception of the USA in the final stages of the Pacific war and of Germany facing the Allied landings in Normandy, the incentives for using gas were weak: neither on the Axis nor on the Allied side had the military commands formed any clear idea of the relative merits of gas as an offensive weapon compared with other weapons. Reluctant for psychological, political and institutional reasons to attach much weight to the claims of the proponents of gas warfare, they left the initiative to the enemy. CW capabilities for which there was little rational justification grew up on either side, generated at least as much by the momentum of past events as by considerations of possible future use. Their growth was perceived in exaggerated form by both sides, and both sides were content to be deterred by them. The war ended with the military establishments still unconvinced that gas was a generally valuable weapon.

5. *The Convention on the Prohibition of Biological Weapons**

The Convention on the prohibition of the development, production and stockpiling of bacteriological (biological) and toxin weapons and on their destruction entered into force on 26 March 1975. Under Article II of the convention, the parties are obliged to destroy, or to divert to peaceful purposes, all prohibited agents, toxins, weapons, equipment and means of delivery in their possession, not later than nine months after the convention becomes effective. Accordingly, the USA stated that 'the entire United States stockpile of biological and toxin agents and weapons has already been destroyed, and our former biological warfare facilities have been converted to peaceful uses'. The United Kingdom said that it had 'no stocks of biological weapons' and that under the legislation in force in the areas covered by the UK ratification, it was now a criminal offence for anyone to be involved in the activities prohibited by the convention. The USSR made the following announcement:

> In accordance with the legislation and practice of the Soviet Union, compliance with the provisions of the Convention on the Prohibition of Bacteriological (Biological) and Toxic Weapons, which was ratified by decree of the Presidium of the Supreme Soviet of the USSR dated 11 February 1975, is guaranteed by the appropriate State institutions of the USSR. At present, the Soviet Union does not possess any bacteriological (biological) agents and toxins, weapons, equipment or means of delivery, as referred to in article I of the Convention.

Verification of the accuracy of such statements has not been envisaged in the convention.

In recent months, some press reports alleged that the US presidential directive on the destruction of biological weapons had been circumvented and that the Soviet Union was building new facilities for the manufacture

* From *SIPRI Yearbooks*, 1972 and 1976.

and storage of biological weapons. Any party to the convention which finds that any other party is acting in breach of obligations deriving from the provisions of the convention may lodge a complaint with the UN Security Council. By 31 December 1975 no such complaint had been submitted.

Significance of the convention

In terms of disarmament, the convention is a preventive measure: it prevents the spread of biological and toxin weapons to countries which do not possess them now; it prevents the development of biological agents militarily more attractive than the existing ones, which may result from scientific advances modifying the conditions of their production, stockpiling and use. But the abolition of the means for biological warfare by those possessing them is also the first real disarmament step taken during the whole post-war period, the only one involving any measure of military 'sacrifice'.

From the legal point of view, the convention strengthens the force of the unilateral renunciations of biological weapons made by a number of nations in recent years. It imposes equal and identical obligations on all.

Last, but not least, it opens new prospects for international scientific cooperation in the field of peaceful uses of microbiology; the cooperation would be enhanced if at least a portion of the savings derived from biological disarmament is directed to that end.

On the other hand, it is unfortunate that a split has occurred in the treatment of chemical and biological weapons. Since the signing of the Geneva Protocol in 1925, both categories of weaponry have been dealt with inseparably in a number of international documents, and have been associated with each other in a single taboo in the public mind. The technical difficulty of drawing a boundary between chemical weapons and biological weapons adds to the artificiality of the division.

Even more regretable is the fact that, in bisecting the traditional chemical-biological unity, priority has been accorded to agents which, because of their uncontrollability and unpredictability, are of little utility and therefore judged to be militarily less important. Biological disarmament is a marginal disarmament measure compared to the banning of chemical weapons.

6. *Chemical disarmament**

The problem of chemical and biological weapons was brought up at the disarmament conferences during the inter-war period. The League of Nations had asked a number of experts to consider the effects of CB weapons.

* From *The Problem of Chemical and Biological Warfare*, Vol. V, SIPRI, 1971, and *SIPRI Yearbooks*, 1972, 1974 and 1975.

The aim was to increase public awareness of the possible dangers of CBW. The report appeared in 1924 and made the following main points: the use of poisonous gases marked the appearance of a terrible weapon; chemical weapons gave an immense superiority to any power with hostile intentions; the possibilities of camouflaging chemical preparedness were very great; and biological weapons were not particularly formidable.

On the initiative of the League of Nations, an international conference to consider the supervision of the trade in arms was convened in Geneva. The conference did not achieve its main purposes but produced as a side result the Protocol for the Prohibition of the Use of Asphyxiating, Poisonous or Other Gases, and of Bacteriological Methods of Warfare. The Geneva Protocol was signed on 17 June 1925.

Chemical and biological weapons appeared in a greatly different technical and political setting after World War II compared with the inter-war period. The change was mainly due to the following factors:

The advent of nuclear weapons, the destructive power of which overshadowed everything else. The rise of nuclear weapons was probably the main reason for the relatively low attention given to CB weapons during the first years of the post-war period.

The rapid development of nerve gas weapons during the 1940's and 1950's. With the discovery of the V-agents, weapons were created which were many times more deadly than earlier chemical weapons.

The developments in microbiology making biological warfare appear, particularly during the 1960's, a much more threatening possibility than it had seemed before the war.

The United Nations first became involved with the question of CBW in 1947 during discussions on how to define mass destruction weapons. On the initiative of the United States, weapons of mass destruction were then defined to include, *inter alia*, 'lethal chemical and biological weapons'. The actual resolution was worked out in the Commission for Conventional Armaments and passed by the Security Council in August 1948.

The UN Secretary-General's report on the effects of CB weapons was released in July 1969. It stated that if CB weapons were ever to be used on a large scale in war, no one could predict how enduring the effects would be and how they would affect the structure of society and human environment; the dangers would apply as much to the aggressor country as to the country attacked and protective measures would not be of much use; the risk of proliferation of CB weapons was considerable.

The controversy over tear gases

The old controversy over whether tear gases, other irritant agents and antiplant agents were covered by the Geneva Protocol flared up again in the United Nations in the autumn of 1969. On the initiative of Sweden a draft resolution was submitted according to which the Assembly would declare that the Geneva Protocol embodies the generally recognized rules of international law prohibiting the use in international armed conflicts of all biological and chemical methods of warfare; further it would be declared contrary to those rules to use any chemical agents of warfare which had direct toxic effects on man, animals or plants, or any biological agents of warfare which were intended to cause disease or death in man, animals or plants and which depended for their effects on their ability to multiply in the person, animal or plant attacked.

The United States strongly opposed the draft resolution and asserted that tear gases, other irritant agents and antiplant agents were not covered by the Geneva Protocol. In the course of the debate the participants used many arguments from the tear-gas debate in the early 1930s. Finally, however, the resolution was adopted with 80 votes in favour, 3 against (the United States, Australia, Portugal) and 36 abstentions. A few months after the tear-gas debate in the United Nations, the British Government took the view that the use of CS (a tear gas used in Viet-Nam) and other such gases not significantly harmful to man in other than wholly exceptional circumstances was not prohibited under the Geneva Protocol. The British statement, made in February 1970, represented a change in the position which the United Kingdom had upheld since 1930.

The United States Government announced a new CBW policy in a statement by President Nixon in November 1969. The three important decisions were: (*a*) unilateral renunciation of biological warfare including disposal of existing stocks of biological weapons; in February 1970 it was made clear that the renunciation also embraced the so-called toxins; (*b*) renunciation of first use of both lethal and incapacitating chemical weapons; and (*c*) submission of the Geneva Protocol to the Senate for ratification.

Verification

During these years the problem of verifying a prohibition of production and possession of chemical and biological weapons has occupied an increasingly large place in the debate. With respect to verification the main issue is that the United States and the United Kingdom consider that verification of non-production of chemical warfare agents, while absolutely essential in their view, is impossible using any method so far suggested. The most common reason put forward is the close dependence of military chemical production on the civilian industry. The Soviet Union and its allies contend that national means of verification, together with the possibility of making complaints to the Security Council, are sufficient for supervising a production ban on chemical agents.

Comprehensive and partial solutions

The parties to the biological convention concluded in April 1972 undertook a commitment to negotiate 'in good faith' with a view to reaching agreement on effective measures for the prohibition of production and possession of chemical weapons.

In a joint communiqué of 3 July 1974, the governments of the USA and the USSR agreed to consider a joint initiative at the Conference of the Committee on Disarmament (CCD) 'with respect to the conclusion, as a first step, of an international convention dealing with the most dangerous, lethal means of chemical warfare'.

A fully comprehensive agreement prohibiting all activities related to chemical warfare agents and weapons, and providing for their total destruction, would be the best solution from the point of view of disarmament. If there is a choice between a partial convention and no convention at all, a partial solution will be desirable if it is meaningful. To be meaningful, a partial agreement would have to prescribe the destruction of a significant part of existing stockpiles, especially of super-toxic agents and munitions, and to proscribe the development and production of all chemical warfare agents, and of weapons, equipment or means of delivery designed to use them, as well as inter-state transfer of the prohibited items. In addition, areas of possible military confrontation should be made entirely free from chemical weapon stockpiles; such disengagement would further reduce the likelihood of chemical warfare. As long as the stockpiles remain, be it in reduced proportions, the possibility of chemical weapons being resorted to in international conflicts will not disappear.

A draft convention submitted by Japan on 30 April 1974 on 'the prohibition of the development, production and stockpiling of chemical weapons and on their destruction' provided for a complete chemical disarmament to be achieved in successive stages. Japan contended that the supertoxic warfare agents (and possibly also mustard-type agents) should be banned first, because adequate verification of nonproduction of these chemicals is now feasible; the prohibition of the remaining agents would be postponed until equally effective verification measures were devised.

Verification is certainly essential to deter treaty violations or to enable their timely detection but its stringency should be commensurate with the importance of the weapons banned. Supertoxic chemicals are the most dangerous warfare agents. Therefore, they require strict control measures to reduce to a minimum the risk to which a state might be exposed in case of breaches. Non-supertoxic agents have less and, in many cases, negligible value as weapons. Consequently, they require considerably less rigid control. There would seem to be no valid reason why a convention could not be made comprehensive right away, at least as regards development and production, and provide for verification measures with different degrees of stringency, depending on the character of the banned agents. Even the tightest control possible could not prevent diversion of some dual-purpose agents to military purposes. In this case, more reliance would have to be placed on the 1925 Geneva Protocol prohibiting the use of chemical weapons which a chemical disarmament convention is meant to strengthen, not replace.

To be of any value, a partial ban would have to cover the components of binary weapons. Otherwise, one category of lethal weapons would be permitted to be replaced by another, perhaps less effective in combat, but more convenient and safer to handle, transport and store.

*7. The case against nerve gas**

During World War II Germany produced, but did not use, a new super-toxic class of lethal chemicals, the nerve gases. During the 1950's and 1960's the United States produced thousands of tons of nerve gas and stockpiled nerve gas weapons designed for tactical battlefield use. These are stored mainly in the continental US, with a lesser quantity deployed in Europe, in the Federal Republic of Germany. We have no reliable estimate of the size of the composition of the Soviet poison gas stockpile, although the USSR and a number of other countries could readily produce nerve gas.

Lethal chemicals are generally considered to be weapons of mass destruction. For example, under not uncommon meteorological conditions a single light-bomber could deliver enough nerve gas to cause a high percentage of fatalities over a downwind area of several square miles. But despite the potential of nerve gas and certain other lethal chemicals for inflicting mass casualties, quite effective protection can be provided for combat troops, in the form of modern gas masks, protective clothing, vehicle air conditioners, and other equipment. Although an initial resort to nerve gas would inflict heavy casualties on military units if caught off guard, its subsequent use against troops with modern protective equipment would be much less effective, a fact of potential importance for chemical arms control.

The Geneva Protocol and proliferation of chemical weapons

The principal treaty dealing with chemical weapons is the Geneva Protocol of 1925. All militarily important nations are parties, including members of NATO, the Warsaw Pact, and the People's Republic of China. After nearly fifty years of alternating controversy and inattention, the United States has finally become a party to the Protocol, following its ratification by President Ford on January 22, 1975, with the undivided support of the Senate. The Protocol is, in effect, a no-first-use agreement. It *does not* prohibit stockpiling of chemical weapons or reprisal in kind against a violator. However, the US and USSR as parties to the Biological Weapons Convention of 1972 have undertaken, under Article IX, to negotiate effective measures for prohibiting the development, production and possession of chemical weapons of war.

* Shortened version of Matthew Meselson: 'What Policy for Nerve Gas?', *Arms Control Today*, Vol. 5, No. 4, April 1975.

Meanwhile the Department of Defense has renewed its request, voted down in the House of Representatives last year, for funds to build a facility to produce a new generation of nerve gas weapons, safer to handle and store, called binaries. The case for buying binaries is not that they are more effective on the battlefield—in fact they are not. Rather, the arguments for and against them are largely psychological and political. Advocates consider that their safety features will overcome public opposition to transportation and forward deployment of nerve gas weapons. Critics argue that a major new round of chemical weapons procurement will spoil chances for negotiating a chemical arms control treaty and will stimulate the international proliferation of chemical weapons.

US policy at a crossroads

While the dispute over binaries has occupied center stage, the present situation represents a crossroads of a more fundamental nature. Broadly defined, the choice is between (1) replacement of the existing stockpile with binaries or at least the retention of the current inventory, possibly with some modifications to suit newer types of aircraft and artillery or, (2) renunciation of lethal chemical weapons, either through international agreement or unilaterally, seeking a treaty afterwards.

It is not maintained that chemical weapons are needed to deter war itself. Our conventional and nuclear forces serve that role. Neither do senior officials consider that we would have any important incentive to be the first to attack with gas should major war occur. Stated US policy has long been not to start poison gas warfare, a doctrine further solidified by US ratification of the Geneva Protocol. Rather, it is argued that the prospect of retaliation in kind would contribute importantly to deterring the Soviets from using the nerve gas that they must be assumed to possess and that, if such deterence fails, our retaliation could enable us to defend Europe without necessitating immediate resort to nuclear weapons. The rationale for these beliefs rests not on the direct casuality-producing capability of nerve gas, which would be minimized by the use of protective equipment, but rather on the reduction in fighting efficiency that results from wearing masks and suits and taking other protective measures. It is contended that the ability to retaliate in kind in the combat zone and in rear support areas would allow us to impose on the Soviets the same protective posture they impose on us, greatly reducing the advantages to them of any protracted use of gas. However, it must be admitted that our retaliatory capability does nothing to reduce the advantage to the Soviets inherent in the initial casualties and confusion that could be inflicted by a surprise gas attack on our forces.

Setting the example

It is generally agreed that in addition to the cost in resources, there are other costs of stockpiling nerve gas and having an active nerve gas program.

Today, no non-nuclear nation is thought to have stockpiled nerve gas weapons. It is very much in our interest to preserve this situation. Our great wealth allows us to expend enormous quantities of conventional munitions in tactical war and to maintain large strategic and tactical nuclear forces. Very few countries even approach this capability. However, nerve gas weapons have the potential of wide area coverage at relatively low cost. Their proliferation would greatly enhance the capability of smaller countries and perhaps even of dissident paramilitary groups for threat, harassment, and destruction. The United States and the Soviet Union set the pace and direction of military developments throughout the world. The more interest we display in nerve gas weapons, the more we pioneer their technology and invest in them, the more lesser military powers are likely to question their case for refraining from acquiring nerve gas weapons of their own.

On a different level of concern, the rapid and accelerating advancement of biochemistry and the biological sciences is inevitably leading to a profound ability to manipulate life processes for good or ill. Over the long run, it may be very important to create an international consensus that such knowledge is not to be exploited for military purposes. The possession of nerve gas weapons maintains institutional commitments to such exploitation. In contrast, if nerve gas can be eliminated we would be free to create an atmosphere in which our increasing knowledge of life processes is directed solely to man's benefit and in which research is conducted under the more or less open public scrutiny that is probably necessary to ensure such beneficial use.

*8. Binary nerve-gas weapons**

In the US President's foreign policy report to Congress of May, 1973, it was stated that the US administration remained 'firmly committed to achieving effective international restraints on chemical weapons'.

But within the US Department of Defense (DoD) the chemical-weapon programme has been proceeding under a momentum of its own. In September 1973 the Secretary of the Army announced plans for the first stage of a new round of chemical-weapon buying. The immediate objective was $200 million of artillery projectiles embodying a novel technique for disseminating standard nerve gases, the so-called 'binary' technique. Binary munitions differ from their predecessors in a safety feature of their basic design: they are loaded with retractable containers of chemicals having only moderate toxicity which react together to produce nerve gas during the final trajectory of the munition to its target. Next in line for procurement after the artillery projectiles is the *Bigeye* aircraft bomb, a joint Navy/Air Force binary development that the Navy began in 1965. Rockets, spray-tanks, cluster bombs and missile warheads are also included in the binary

* From *Chemical Disarmament: New Weapons for Old*, SIPRI, 1975.

R&D programme, currently funded at about $10 million per year. DoD officials have cited an ultimate objective of replacing the entire US nerve-gas stockpile with binary munitions.

A foreign-policy collision course

Work on binary munitions has been continuing in the United States for some 20 years, but the programme has become a concerted one only within the past ten years. It would be a mistake to suppose that anything like a complete nerve-gas capability, exploiting the binary concept, is now accessible to procurement. That situation is still a long way off, in fact well over a decade away, or probably two decades, at the present rate of progress. What is now nearly available is a limited and unproven binary capability for two particular artillery weapons—the 155-mm and 8-inch howitzer systems. These weapons, it is true, have always been considered to offer one of the better ways of spreading nerve gas in combat situations; but on the evidence of existing nerve-gase stockpiles, DoD authorities seem to believe that a nerve-gas capability cannot adequately fulfil its stated mission for the United States unless it comprises a much wider range of weapons.

The most difficult requirement made of the chemists working in the binary programme is that they should provide pairs of low-toxicity chemical reactants that are, on the one hand, capable of spontaneously reacting together in a space of seconds to produce a high yield of nerve gas, and on the other hand, sufficiently stable to survive prolonged periods of storage. But highly reactive chemicals tend to be unstable on storage; they also tend to react with biochemical substances in the human body, thus exerting toxic effects. Yet another difficulty is that the chemicals should ideally be liquid at ambient temperatures: the involvement of solid chemicals complicates the mechanics of binaries.

The binary programme seems to be another instance in which the DoD is becoming locked onto a foreign-policy collision course by its own standard operating procedures. The lock in this case is the dogma of like-with-like deterrence—the unquestioned belief that a retaliatory nerve-gas capability is the best safeguard against the possibility of an attack with chemical weapons on NATO forces in Europe. The driving force is the notion of credibility, a concept whose definition and application can satisfy an important constituency.

Assimilation of nerve gas into the armed services

For the time being chemical warfare remains an unconventional, unaccepted, illegal method of warfare, held in check by a complex array of constraints and inhibitions. To loosen a part of the array could eventually bring down the barriers completely. It is in the interests of the United States, and those of everyone else, to prevent yet another implement of mass destruction from becoming accepted and assimilated into the arsenals.

The dangers here are by no means theoretical ones. They may be seen as a practical problem of command and control. Just as binaries may facilitate nerve-gas stockpile-management in rear areas, so also may they simplify logistical procedures in forward areas. The stringent precautions which at present surround, and complicate, nerve-gas supply may come to appear superfluous because of the safety features of binaries. Pressures may develop to move them out of the 'special ammunition' supply category which nerve gas at present shares (in France, as in the United States) with nuclear munitions, thus enabling binaries to be stored further forward in combat zones. Filled or semifilled binary munitions might ultimately become part of the basic load of tactical forces. With each successive relaxation of controls, the possibility of inadvertent or unauthorized initiation of chemical warfare would be substantially increased. By accelerating the assimilation of nerve gas into the armed services, binaries would also increase the likelihood of it being used.

9. *Delayed toxic effects of chemical warfare agents**

Only very few specialists today know a great deal about the variety and extent of delayed lesions caused by chemical warfare (CW) agents and other militarily important chemicals. The military experts among them, and the representatives of industries manufacturing these poisons, have shown little interest in disseminating the findings on delayed lesions. Nevertheless, some of this information has managed to reach the public, which has been deeply aroused by the use of CW agents by the USA in Viet-Nam.

The term *delayed lesion* is defined to mean a lesion—caused by acute or subacute poisoning by CW agents—noticed either as a residual injury or by the unexpected onset of related symptoms after a protracted period of months or years, and being essentially irreversible. Effects of delayed lesions may be summarized as follows:

1. In addition to causing acute effects, most CW agents are liable to cause delayed effects. This applies to persons who have survived acute poisoning, as well as to those who have sustained subacute—even imperceptible—poisoning.

2. The main delayed effects take the form of (*a*) psychopathological-neurological changes; (*b*) malignant tumours (cancer); (*c*) increased susceptibility to infectious diseases (primarily of the lungs and upper respiratory tract); (*d*) disturbances of liver function; (*e*) pathological changes in the blood and bone marrow; (*f*) eye lesions; and (*g*) premature decline in vigour, rapid aging and related functional disturbances, such as decline in potency and libido.

* From *Delayed Toxic Effects of Chemical Warfare Agents*, SIPRI, 1975.

3. In addition to these neural and organ-specific effects, mutagenic effects in human beings—although not conclusively proven—must be taken into account. The occurrence of teratogenic and embryotoxic effects in human beings—likewise not proven in all respects—is highly probable.

4. What has been said of CW agents also applies to a large number of compounds related to them in structure and action. These compounds are of scientific and practical interest either as by-products or as 'model' compounds.

5. The biochemical mechanism of the delayed effects of most CW agents and related compounds is not yet known. In the few cases where concrete information is available, research work is still incomplete.

6. No effective prophylactic or therapeutic measures are available, at present, against the delayed effects of CW agents. This means that the handling of these poisons—no matter how slight—in the course of their manufacture, storage or use carries the risk of imperceptible poisoning and can lead to any of the above-mentioned types of related delayed lesions.

*10. Chemical-warfare policy determinants**

For the European battlefields of World War I, where massive concentrations of troops had been forced by the new 'high' explosives and the machine gun into the static confines of opposing trench systems, chemical weapons provided a form of firepower that was well suited to prevailing military doctrine. But the concurrent development of the tank and the aircraft, in particular, was subsequently to bring about a profound realignment of the firepower-mobility-dispersion triad which dictates military doctrine, thereby imposing new doctrines which could accomodate the chemical weapons of the period only with difficulty. As a consequence, military interest in poison gas languished (except, in some countries, as a weapon against primitive opposition), and by the end of World War II it was viewed more as an awkward remnant of history than as a useful implement of modern warfare, now likely to be dominated by nuclear weapons.

Yet after the war, the victorious Allies discovered that a new family of chemical-warfare agents—the first of the nerve gases—had reached an advanced state of development in Germany. In the West, and no doubt also in the USSR, it soon became apparent to CW specialists that the offensive properties of the new chemicals could be harnessed in weapons that would be tens or hundreds of times more powerful than their predecessors, and which would be less unsuited to a warfare of rapid mobility and wide dispersion. Even though military authorities still remained unenthusiastic

* Excerpts from Julian Perry Robinson: 'The United States Binary Nerve-Gas Programme: National and International Implications', *ISIO monographs*, Institute for the Study of International Organisation, University of Sussex, 1975.

about chemical warfare, the nerve gases provided a new lease of life for the chemical-weapons R&D establishments. A new momentum began to build up in the chemical-warfare programmes, impelled by the deepening mistrusts of the Cold War. The Korean War afforded scenarios with which the protagonists of chemical weapons could assert the military utility of nerve gas (most notably as means for countering 'human wave' infantry assaults), and by 1954, as a consequence of across-the-board increases in military appropriations, agent GB (sarin nerve gas) was in full-scale production in the United States. The US Army Chemical Corps, which had inter-Service responsibility for CW supplies, initiated an active campaign to broaden Service awareness of its commodities and to re-interest military planners. At a time when military thinking was preoccupied with nuclear weapons and the doctrine of 'massive retaliation', this was no easy task. It was facilitated, first, by the discovery in England of a new family of nerve gases, the still more potent V agents, and later on by the shift in strategic doctrine towards the concept of 'flexible response'. The Congress was persuaded to finance the construction of another nerve-gas factory, for agent VX (opened in 1961), to treble Chemical Corps R&D appropriations, and to step up chemical-weapons procurement funding. Thus, after a period of regression at the time of World War II, chemical weapons were once more advancing towards acceptance and assimilation by the armed services of at least one world power.

The public record does not yet reveal the full extent of Allied involvement in this American enterprise. Apart from France the United States is the only Western country that possesses militarily-significant supplies of chemical weapons. On the research side, it is certainly the case that cooperation with Canada and the United Kingdom, in particular, has been very extensive. A recent British official history remarks that after the war 'the programmes of (the UK and the USA) remained so closely in step as to be virtually integrated'; and, in private, senior US Army Chemical Corps personnel speak of British expertise being largely responsible for those fundamental studies in the fields of operations research, agent selection and weapon concepts on which the present US nerve-gas capability is based. What is not known, however, is the degree to which Allied forces are trained, or would be willing, or would be permitted, to employ US chemical ammunitions. Noteworthy in this connection is the statement in the 1970 West German White paper on Defence that 'the Federal Republic has made no preparation for using (chemical weapons), does not train military personnel for that purpose, and will abstain from doing so in the future'. Thus, even though those stocks of chemical weapons which the United States maintains in Europe are all located in West Germany, the Federal Republic has disavowed any intention of seeking access to them in the event of a European war. Nothing at all has been said in public about Ally-access to French chemical weapons.

Still more obscure is the Soviet attitude towards chemical weapons. There is ample evidence in the Soviet military literature that the threat of chemical warfare is taken very seriously, and, as in NATO, chemical warfare protective equipments are widely deployed. But, on the weapon side, in the absence of any official Soviet statements whatsoever, one can only assume that the technological impulse which has driven Western programmes has also been manifest in the USSR. Western Intelligence on the matter is known to be weak. The estimates of the 1950's and 1960's to the effect that 15 per cent of Soviet munitions supply was chemical, and that the Soviet chemical stockpile was 7–10 times larger than the American one, have since been discredited. Even so, Western chemical defence authorities remain convinced that the USSR possesses a powerful nerve gas capability; and by the traditional, if dubious, process of inferring intentions from capabilities, Soviet nerve gas is considered a significant threat to NATO.

Such an assessment presupposes that the USSR and/or her allies will be prepared to abrogate the no-first-use commitments assumed upon ratifying the Geneva Protocol. Alternatively it presupposes that the USSR believes NATO's commitments in this regard to be suspect. This is the basic dilemma of current chemical-warfare policy making. The laws of war, of which the Geneva Protocol is part, derive what strength they have from the sanction of reprisal, so that an obvious policy option—albeit a flawed one—is to maintain a retaliatory nerve-gas capability as a form of deterrent. This indeed is current US policy, and may well be Soviet policy also. The fact that the retaliatory capability is also available for initiation of chemical warfare is no doubt seen as an advantage by those who value a wide range of military options; but it is also seen as a disadvantage, and a compelling reason for negotiating further CW treaty limitations, by those who see in this ambivalence a source of international mistrust and tension.

Clearly there are parallels to be drawn with nuclear weapons, but there is also a fundamental difference. States that possess nuclear weapons have by now been forced to condition themselves into at least some degree of readiness to fight a nuclear war. So, with greater or lesser reluctance, nuclear weapons have been accepted into military doctrine and the doctrine has been modified, via deterrence theory, to accommodate them. Chemical weapons have not yet achieved this degree of assimilation. They therefore lack full credibility either as a threat or as a deterrent. All that can be said with any confidence is that nerve gas is potentially a more powerful casualty agent than any conventional weapon; that its destructiveness and inherent escalatory nature makes it comparable with 'tactical' nuclear weapons; that protection against its effects can in theory be provided to a high degree, but in practice—at present levels of preparedness—to a much lower degree; that if used on a battlefield it would complicate, and probably slow down, the conduct of operations beyond all recognition; and that, technologically and economically, it is accessible to a wider range of States than at present possess it. These are the parameters which must be taken into full consideration in determining chemical-warfare policy. They are therefore also the primary points of reference against which the binary programme must be assessed.

*11. Status of the 1925 Geneva Protocol on CB warfare**

On 17 June 1975, 50 years had elapsed since the signing in Geneva of a protocol for the prohibition of the use in war of asphyxiating, poisonous and other gases and of bacteriological methods of warfare. The agreement was prompted by the shocking experience of World War I during which at least 125 000 tons of toxic chemicals were used and the toxic gas casualties numbered as many as 1 300 000.

The historic significance of the Geneva Protocol lies in the fact that an international legal constraint, 'binding alike the conscience and the practice of nations', was imposed on acts which were generally held in abhorrence and had been condemned by the opinion of the civilized world. Its weakness, however, is the same as that of other laws of war: rules of conduct which are set for belligerents in time of peace may not resist the pressure of military expedience generated in the course of hostilities. Indeed, since 1925 chemical weapons have been used on several occasions. But on each such occasion, the extent of worldwide indignation and censure testified to the immutability of the generally recognized standard of international law, as embodied in the Geneva Protocol. It is, in great part, due to the Protocol that the history of chemical warfare since World War I has been one of relative restraint and that no bacteriological weapons have been used in modern times. Nevertheless, the danger that the weapons prohibited by the Geneva Protocol may, under certain circumstances, be resorted to will not disappear as long as they exist in the military arsenals of states.

During the past 50 years, new, more toxic compounds than those employed in World War I have been discovered, and the means of their dispersion considerably improved. Were these new weapons ever to be used on a large scale, they would cause a tremendous loss of human life, much greater than ever before, with civilians being even more vulnerable than the military; the whole structure of society and the environment in which we live could be affected. Only a complete cessation of the development and production as well as the destruction of the existing stockpiles of the weapons in question could remove this danger. The first step in this direction has already been made with the signing of the biological disarmament convention. However, an agreement on the prohibition of the development, manufacture and stockpiling of chemical weapons, which, from the military point of view, are more important than biological weapons, is still pending in spite of years of international negotiations. In the meantime a new danger has arisen, that of the possible introduction of the so-called binary nerve-gas munition. This is a munition filled with two or more non-toxic chemicals that mix and react when the munition is delivered to the target, the reaction product being a nerve gas. Once these weapons are deployed, the difficulties

---

* From *SIPRI Yearbook*, 1976.

experienced at the chemical disarmament negotiations with regard to verification of compliance may increase to the point of making an agreement well-nigh impossible. Speedy measures in this field are, therefore, urgently needed. But they should not hamper further action aimed at the strengthening of the Geneva Protocol.

With the recent ratification by the USA, all militarily important states are already bound by the Geneva Protocol, but many states are still missing; only 94 nations are party to it. (For the list of non-parties, UN and non-UN, see list below.) And yet, chemical warfare is more likely to occur between small countries than among the great powers. Universal adherence to the Geneva Protocol, as has been repeatedly recommended by the UN General Assembly, would reinforce it considerably. In addition, the parties should accept the application of the protocol to *all* armed conflicts.

A formal reaffirmation by individual states of the comprehensive nature of the ban under the Geneva Protocol would also seem desirable. It would have to be made in accordance with the 1969 UN General Assembly resolution stating that the protocol prohibitions apply to the use of *all* biological and chemical methods of warfare, regardless of any technical developments.

Furthermore, the reservations attached to the protocol by a number of states and limiting its applicability to nations party to it, and to first use only, should be withdrawn to make the prohibitions more universal and absolute. In any event, the reservation concerning the right to use bacteriological methods of warfare against non-parties, or in retaliation, is clearly incompatible with the convention for the complete elimination of biological weapons, and should be declared null and void.

And, finally, the effectiveness of the Geneva Protocol would increase if an international procedure were agreed upon to verify allegations of breaches. Past experience has clearly demonstrated the need for such a procedure.

The total number of actual parties to the Geneva Protocol, as of 31 December 1975, is 94.

List of non-parties (UN members and such non-UN members which are bound by at least one of the post-war arms-control agreements):

Afganistan, Albania, Algeria, Bahamas, Bahrein, Bangladesh, Barbados, Benin (Dahomey), Bhutan, Bolivia, Botswana, Burma, Burundi, Byelorussia, Cambodia, Cape Verde, Chad, Colombia, Comoros, Congo, Costa Rica, Democratic Yemen, El Salvador (signed but not ratified), Equatorial Guinea, Gabon, Grenada, Guatemala, Guinea, Guinea-Bissau, Guyana, Haiti, Honduras, Jordan, Korea South, Laos, Mali, Mauritania, Mozambique, Nicaragua (signed but not ratified), Oman, Papua New Guinea, Peru, Qatar, San Marino, São Tomé and Principe, Senegal, Singapore, Somalia, Sudan, Surinam, Swaziland, Ukraine, United Arab Emirates, United Republic of Cameroon, Uruguay (signed but not ratified), Viet-Nam South, Western Samoa, Zaire and Zambia.

4. Environmental warfare, ecocide, and weapons of mass destruction

1. *Environmental modification—new weapon of war**

The term environmental modification embodies many types of deliberate or inadvertent changes in the environment in which man lives. One particular aspect of this, the deliberate geophysical modification, and its use of waging war, is considered here. Geophysical modification could conveniently be divided into four sections, namely, weather and climate modification, modifications of the oceans and earthquakes, modifications of certain types of electromagnetic radiation reaching the earth and modification of certain electrical behavior of the atmosphere. Most such modifications are based on the fact that nature has stored, in some regions of the earth and its surroundings, far greater amounts of energy than usual, causing instabilities. If such environmental instabilities are identified and triggered by the addition of a small amount of energy, a considerably greater quantity of energy would be released.

Weather and climate modification

The atmosphere is a very complex physical and chemical system. The complexity arises particularly from the fact that it is a dynamic system. The phenomena of weather and climate result from the interaction of atmospheric processes, which range from large-scale processes like the air circulations round the earth caused by the differences in radiation arriving from the sun in equatorial and polar regions, to small-scale processes such as the exchanges of water molecules on the surfaces of minute drops or on ice crystals in clouds. A considerable amount of research has been carried out recently in order to use weather and climate modifications as weapons of war. Since considerable amounts of energy are required to modify weather by altering the dynamic processes, it is usual to change the weather by altering the microphysical processes. This is achieved by introducing into the clouds materials such as water droplets, dry ice, solid CO_2, silver and lead iodide or

* Shortened version of a paper by Bhupendra M. Jasani from *Ambio*, Vol. 4, No. 5–6, 1975.

liquid propane; seeding agents which change the characteristics and the nature of the cloud particles.

It is possible to modify rainfall or snowfall only under certain conditions. It has been suggested that rain or snow modification could be used in warfare as either a tactical or a strategic weapon. In the former case it might be used as a direct weapon. Increasing the rainfall, for example, could interfere with the movement of troops and supplies. Increasing the snowfall by cloud-seeding in mountainous areas could make transportation and communications in such areas more difficult. As a strategic weapon, rain could be modified by cloud seeding over a long period over one's own country, so that rainfall could either be increased or decreased in neighbouring states when the seeded clouds pass over them.

However, such controlled precipitation is not always possible and successful, as seen in Vietnam.

Weather modification does not constitute the modification of rain only but also fog, hail, severe storms, cloud electricity and lightning modifications.

Modification of severe storms

The interest in the control of severe storms arises from the fact that they are among the most destructive of all natural phenomena, causing enormous amounts of damage to property and lives. There are two types of such storms, tornados and hurricanes.

Tornados are the most violent of these storms but they are small, do not last long and the destruction caused by them is confined to a narrow track. The energy stored in tornados is, however, very large, equivalent to some 50 kt of TNT. They consist of violently rotating columns of air in contact with the ground and they also produce hail and lightning.

Very little is known about tornados at present, and therefore, not many attempts have been made to modify them. Some possible ways of modifying them, however, have been suggested. One method consists of modifying the flow of wind close to the surface of the earth and another method consists of altering the precipitation processes by seeding.

Of the two types of storms, hurricanes are the most destructive. Energy stored in a hurricane could be of the order of 1 000 megatonnes of TNT. Hurricanes are formed over warm tropical waters and begin dissipating soon after moving over either cool water or land.

Theoretically, it is possible to use hurricanes and other storms as weapons by enhancing, dissipating or guiding them by means of cloud seeding or other techniques. A controlled hurricane could be used against a country with extensive coastlines. A basic requirement is that the atmosphere stratification is potentially unstable to the same degree as the tropical atmosphere. In order to generate a hurricane, heat transfer from the sea to the atmosphere is essential. The evaporation from large volumes of water can be modified by spreading thin layers of materials, such as oil, over the water surface. This technique, together with cloud seeding, could, in theory, be used to guide a hurricane to destroy coastal defences of the enemy. However, attempts at storm modifications made so far have not yielded positive results, so the use of such phenomena as war weapons appears to be purely speculative.

Climate modification

Climate is a result of very complex physical, chemical and dynamical processes and their interactions in the ocean and atmosphere and at the land surface. In order to understand the interaction of these, models of processes leading up to the formation of climate have been constructed. During the past two decades or so, attempts have been made to understand the climate with the use of these models and high speed computers. The results indicated that with levels of thermal energy produced by man today, there would be only minor modification of the earth-atmosphere heat balance, causing very limited change in the climate.

Modification of climate is, on the whole, still in the realm of theoretical possibility, but it is envisaged as a strategic weapon which could, for example, be used to destroy the enemy's agricultural pattern. A number of methods have been suggested above which could be used as triggering mechanisms for climate modification. However, it is only when these triggering mechanisms start a number of other processes in a predetermined way in the general circulation pattern in the atmosphere that climatic modification may result. And this is where the difficulty lies. All the atmospheric processes are not clearly understood, so they cannot be predicted. It is thought that it may not be possible to achieve such changes in large-scale atmospheric circulation in the coming two or three decades.

Manipulation of certain electromagnetic radiation

A considerable amount of discussion is taking place at present on the biological effects of electromagnetic radiation in the region of wavelengths shorter than 300 nm (1 nm= 10^{9} meter). This is the wavelength region of ultraviolet radiation. If more ultraviolet radiation than the present level reaches the earth, it would have adverse effects on all biological life. The reduction of the levels of certain gases in the atmosphere could result in such an increase; the reduction of the level of ozone would produce the greatest increase.

Ozone is located mainly in the stratosphere at an altitude ranging from 10 to 50 km. It is a minor but an extremely important constituent of the earth's atmosphere. The small amount of ozone is essential to protect the life on earth from lethal ultraviolet (uv) radiation.

It is worth mentioning here a few effects of increased uv radiation on biological systems. Skin cancers are caused by exposure to intense uv radiation. Photosynthesis of plants is inhibited and their growth is reduced. Some plants can even die due to the action of the uv radiation. There is little doubt that if a reduction in the concentration of ozone to 50 per cent of its present value could be achieved, then it would have far-reaching effects on the biological systems on earth.

There are two methods available for modifying the ozone layer. One is to use an ozone reactive chemical to reduce the amount of ozone from a small area, thereby exposing a small region on the earth to intense uv radiation. Another technique is to use nuclear explosions within the ozone layer to make a 'hole' in it above the enemy territory.

It is clear that man has within his reach the ability to cause large reductions in the content of the atmospheric ozone but whether he can use it as a weapon will depend on his precise knowledge of the diffusion rate of ozone into the depleted area and the wind conditions in the stratosphere. Both these factors determine the time for which the region below on earth is exposed to ultraviolet radiation. Moreover, the ozone layer is not a well defined layer, but there is a vertical distribution of the atmospheric ozone, and therefore, it may not be so easy to create a small ozone-depleted area.

Modification of the oceans and earthquakes

The properties of the atmosphere are not well understood. The mechanisms of oceans and earthquakes are even less well understood. Very limited attempts have been made to modify earthquakes and only theoretical suggestions have been put forward as to how the oceans could be modified.

The behavior of the oceans is still not fully understood, but during the past two decades or so methods have been devised for predicting the surface wave and surface wind distribution. Some knowledge about certain ocean currents is available, but their variability is not fully known. Therefore, instabilities that may exist within the oceanic circulation and which may be manipulated have not been identified, so any discussion on the modification of the ocean currents is largely speculative.

Another such area on which speculations have been made concerns tsunamis, an extremely violent type of tidal wave. These often result when sediments and rocks perched on the continental shelf fall or slide into the deep ocean, or occur as a result of earthquakes. Vast quantities of energy are released by the movements of such sediments and rocks. A series of phased explosions could be used to create such movements.

Another method of creating tsunamis may be to use nuclear explosions, either under water or along the base of a large ice sheet, causing it to slide outward into the water. If large ice velocities are achieved, this could create tsunamis causing enormous damage to coastal regions.

Modification of certain electrical behavior of the atmosphere

A more exotic form of geophysical modification is that of the electrical behavior of the atmosphere. There exists an ionized region high in the earth's atmosphere called the ionosphere which extends from about 50 km to hundreds of km above the earth's surface.

Both the ionosphere and the earth's surface conduct electricity in such a way as to cause, e.g., radio waves to be reflected. This phenomenon is used in long-distance radio communication, where the radio waves are reflected back and forth between the ionosphere and the ground. If, however, the ionosphere is modified, for example by means of a nuclear explosion, radio communication could be hampered.

Another use of this ionosphere-earth wave guide may be to propagate very low frequency radiation through it in such a way that this may possibly influence the behavior of individuals through the interaction of this radiation with the electrical activity of the brain.

The dangers ahead

At present the use of geophysical techniques of warfare poses a number of problems. Various techniques employed for weather modifications are useful only under certain meteorological conditions which occur only at certain times of the year and only in certain areas. None of these variables would be completely controllable by man.

One of the disturbing aspects of geophysical modification is that such operations could be carried out covertly. A state could seed clouds over its own territory, knowing that this could cause changes in the rainfall or snowfall over the neighboring state. The state downwind could attribute such changes to natural fluctuations. At present it is not always possible to determine sufficiently accurately whether or not changes in the weather and climate are caused by man. Under such circumstances, the development of such techniques as weapons of war could only increase the conflict and tensions arising from such disasters as crop failures resulting from droughts or floods.

Should the development of the technology continue for its use in war, it will be very difficult to check its proliferation.

2. *Prohibition of environmental means of warfare**

On 21 August 1975, after a series of secret bilateral talks, the USA and the USSR simultaneously submitted to the Conference of the Committee on Disarmament (CCD) identical draft conventions 'on the prohibition of military or any other hostile use of environmental modification techniques'. The convention would prohibit military or any other hostile use (but not the threat of use) of 'environmental modification techniques'. The latter term was defined as any technique for changing—through the deliberate manipulation of natural processes—the dynamics, composition or structure of the

* From *SIPRI Yearbook*, 1976.

earth, including its biota, lithosphere, hydrosphere and atmosphere, or of outer space.

The draft convention listed the effects which could be caused by the use of the prohibited techniques. These were: earthquakes and tsunamis, an upset in the ecological balance of a region, or changes in weather patterns (clouds, precipitation, cyclones of various types and tornadic storms), in the state of the ozone layer or ionosphere, in climate patterns, or in ocean currents. The list was illustrative but the choice of examples seemed haphazard.

To estimate the feasibility of influencing the environment in such a way as to produce the mentioned effects, Canada submitted a working paper identifying the conceivable techniques. It grouped 19 environmental modification techniques within three main categories: atmospheric modification, including the high atmosphere and ionosphere, modification of the oceans, and modification of land masses and water systems associated with them. They were as follows: (1) fog and cloud dispersion; (2) fog and cloud generation; (3) hailstone production; (4) release of material which might alter the electrical properties of the atmosphere; (5) introduction of electromagnetic fields into the atmosphere; (6) generating and directing destructive storms; (7) rain- and snow-making; (8) control of lightning; (9) climate modifications; (10) disruption of the ionized or ozone layers; (11) change of the physical, chemical and electrical parameters of the seas and oceans; (12) addition of radioactive material into the oceans and seas; (13) generation of large tidal waves (tsunamis); (14) stimulation of earthquakes/tsunamis; (15) large-scale burning of vegetation; (16) generation of avalanches and landslides; (17) surface modification in permafrost areas; (18) river diversion; and (19) stimulation of volcanoes.

The stated purpose of the draft convention was to 'limit the potential danger to mankind from means of warfare involving the use of environmental modification techniques'. The aim was not to eliminate the danger altogether by banning *all* environmental modification techniques for military or other hostile purposes. Only those which have 'widespread, long-lasting or severe effects' would be prohibited. The draft convention would only prohibit the environmental modification techniques 'as weapons', leaving aside the environmental impact of other weapons.

What is not explicitly prohibited may be taken as implicitly permitted. Consequently, techniques which do not have widespread, long-lasting or severe effects would be exempted from the ban. Moreover, the meaning of 'widespread, long-lasting or severe', qualifying the prohibition and reducing its scope, was not explained.

The provisions of the draft convention could be interpreted as not covering, for example, fog and cloud generation and dispersal, hailstone production, rain- and snow-making, or increasing the intensity of lightning discharges, because of the limited effects they may have. And yet, these techniques are the most feasible methods to influence the environment for hostile purposes. Such loopholes would disappear if *all* techniques for modifying the environment for military or other hostile purposes were outlawed; no enumeration of the prohibited activities would then be needed. An exemption, however, could be made for techniques which, while used for military purposes, do not produce effects beyond the borders of the state applying them. For example, the generation or dispersal of fog over one's own airfields and ports would remain permitted.

The ban should not be limited to nations party to the convention, as was the case in the US–Soviet draft. Because of the very nature of environmental modification, it may be very difficult to circumscribe its effects geographically. No nation could feel immune to large-scale changes in the environment, wherever they happen to take place.

Research and development

Another gap in the draft convention was the lack of prohibition on research and development of environmental modification techniques for warlike purposes. The reason the sponsors gave for this omission was that verification would be difficult because of the 'dual applicability to civilian and military ends of much research and development in this field'. But not all peaceful modification activities overlap with military; it would be difficult, for example, to envisage the generation of tsunamis for peaceful purposes. Furthermore, research and development in the environmental field could be placed under strict civilian control, thus minimizing the possibility of abuse by the military. Also an open, wide and institutionalized exchange of relevant scientific information, including an obligation to register and place under international observation all major peaceful experimentation with environmental modification techniques, might help to increase confidence among the parties that the obligations are being observed.

The US–Soviet draft convention is a proposal for non-use of certain methods of warfare. It cannot be considered as a disarmament or arms control measure because it does not envisage elimination of a specific weapon from the arsenals of states or prevention of its acquisition. Experience has shown that a non-use commitment, contracted in time of peace, may not resist the pressure of expediency generated in time of war. A ban on the very possession of a given weapon or warfare technique, including research, or at least development, can provide a more reliable guarantee of non-use.

The incentive to develop these techniques is great. It has been estimated that in the USA alone, the average annual costs from damage that could be directly identified with hurricanes, tornadoes, hail, lightning and fog exceed $2 billion, not to speak of agricultural losses due to drought, which could be enormous. In the field of peaceful application of environmental modification, much is already being done on the national level. As regards international activities, the World Meteorological Organization (WMO) has been conducting studies of artificial weather modification, including the enhancement of precipitation, hail suppression, fog dispersal and reduction of wind speed in tropical cyclones. It should be added that the WMO is responsible for the international planning and coordination of the world weather monitoring system which could be an essential part of any system for monitoring large-scale weather- and climate-modification operations. Also the United Nations Environment Program (UNEP) is involved in environmental modification problems. Further internationalization of research and development in the field of environmental modification for peaceful purposes would, apart from obvious scientific, economic and technological advantages, provide some reassurance that substantial resources were not diverted to military ends.

Separating fact-finding and political judgement

Under the draft, any state which finds that another state is acting in breach of obligations deriving from the provisions of the convention would have the right to turn to the UN Security Council and lodge a complaint. The complaint should include all possible evidence 'confirming its validity'. The Security Council would determine whether the allegation was properly substantiated and deserved consideration. Possible investigations could be initiated only by the Security Council, which would also be the sole body authorized to evaluate their results and decide whether a party had been 'harmed or is likely to be harmed' as a result of violation of the convention. The provision of assistance to the victim 'which so requests' would depend on these findings.

The draft required that the complaints procedure should be carried out 'in accordance with the provisions of the Charter of the United Nations'. Since the Charter provides that Security Council decisions on substantive matters should be made by an affirmative vote of nine of the 15 members of the Council, including the concurring votes of its permanent members—China, France, the UK, the USA and the USSR—each important step in the process of verifying allegations of breaches of the convention could be blocked by a negative vote of the great powers. There can be no doubt that this right of veto would be taken advantage of whenever an accusation were directed against any of these powers or their allies. Considering that at the present time, and probably in the foreseeable future, the most likely offenders are precisely the great powers, or some of their allies, which are the only states engaged in large-scale research and development of environmental modification techniques, the verification provisions, as formulated in the draft, were devoid of practical significance.

No treaty can change the prerogatives of the permanent members of the Security Council as long as the UN Charter remains unchanged. But there is, perhaps, no need to involve the Security Council in the implementation of a convention concluded outside the framework of the UN. If, nevertheless, it is thought advisable to make use of the UN machinery and to resort to the services of the Security Council in view of its responsibility for the maintenance of international peace and security, there appears to be no reason why a single body, whatever its standing, should combine the power of conducting investigations with that of determining the guilt or innocence of states with regard to the observance of a treaty. Separating the fact-finding duties from political judgement could render the verification provisions more plausible.

Constraints on new weapons before they have been fully developed, and especially on warfare techniques which are inherently indiscriminate and unpredictable in their effects, could, as preventive measures, contribute to the circumscription of the arms race. But to be effective, the constraints must be comprehensive and contain no loopholes. The US–Soviet draft convention on the prohibition of environmental modification techniques did not meet the above requirements. It would ban the use of these techniques, without banning their development. Moreover, even the non-use commitment was qualified. It was limited to those techniques which produce widespread, long-lasting or severe effects, and which, because they are of uncertain effectiveness, unpredictable and double-edged, that is, potentially hazardous to the user himself, can hardly be conceived as weapons of war. On the other hand, the techniques which do not produce widespread, long-lasting or severe effects, but which could be important in tactical military operations because of their ability to hit more precisely a selected area, would escape the ban. Another important drawback was the lack of an impartial machinery to establish facts of violation. The proposed complaints procedure depending entirely on the good-will of the permanent members of the Security Council seems to be of little value. Elimination of these, as well as other shortcomings described above, could make the contemplated agreement really meaningful.

*3. Proscription of ecocide**

It is axiomatic that warfare is detrimental to the environment. To begin with, the preparation for war is detrimental in several ways. It consumes scarce and nonrenewable resources, usually on a priority basis. It also generates a variety of pollutants during the manufacture and testing of the war material, thereby having an adverse effect on still other natural resources. The present arms race is particularly wasteful of our global resources. Second, the practice of war is detrimental to the environment. Military action consumes nonrenewable resources, often at a lavish rate. And, of course, the battles debilitate the theater of war itself.

There are, however, a number of reasons for examining the ecological dimensions of warfare in some detail. First, the subject has been almost universally ignored in the past, precluding a rigorous assessment of its importance. Second, world environmental problems appear to have reached the point where any substantial regional perturbation of the physical environment becomes a matter of international concern. Third, certain methods of

* Shortened version of a paper by Arthur H. Westing from *The Bulletin of Atomic Scientists*, Vol. XXX, No. 1, January 1974.

warfare likely to be practised in the future with increasing regularity are particularly destructive of the environment.

Lesson of Indochina war

Although the United States seems to have made every effort to control as large a proportion of the Indochinese population as it could (i.e., to win as many hearts and minds as possible), at no time during the war did it ever attempt to physically control more than a minute fraction of the land area of Indochina. In fact, despite occasional US infantry operations, the so-called search-and-destroy operations, essentially no real estate changed hands during the entire war.

What was carried on over the years with a remarkable amount of vigor was the attempt by the United States to make a major fraction of Indochina's land surface continuously inhospitable to its enemy. This had the dual purpose of driving the indigenous rural peoples into the US-controlled population centers and of depriving the enemy combatants of local support (willing or otherwise). One might add that it had the additional effect of making these displaced persons dependent upon US largesse. This strategy of counterinsurgency warfare has been referred to as 'forced-draft urbanization'.

The United States carried out its grand strategy of widespread area denial in a number of suitably grand and innovative fashions. These need only to be mentioned here since they and their devastating ecological impact have been described adequately elsewhere. Perhaps the most bizarre means of denying forest cover and sanctuary to the enemy was via the aerial spraying of vast areas with plant-killing poisons or herbicides. However, forested regions that were particularly troublesome from a military standpoint were destroyed with even greater finality, if less finesse, simply be being scraped away by companies of massed giant tractors (Rome plows). But the most devastating and least appreciated approach to area denial on a continuing basis was the program of pattern or saturation bombing, truly awe-inspiring in its magnitude. Among the other counterinsurgency activities by the United States in Indochina of potential ecological significance were the weather modification and the forest fire (fire storm) programs.

The precise over-all extent of ecological damage to Indochina may never be known. It is clear, however, that in the process of attempting to deny its enemy freedom of movement and local support in the rural and wild reaches of Indochina, the United States has significantly debilitated the forests, the wildlife, the soils and the very ecosystems via its lavish employment of area denial weaponry. Such widespread ecological debilitation becomes particularly serious to a largely agrarian society that must depend to a great extent on an adequate natural resource base for its well-being. Indeed, one of the saddest results of the war has been the separation of a peasant people from its land.

It can be seen that the Indochina war has demonstrated the strategic attractiveness and efficiency to the military of ecologically destructive techniques. It has thereby imposed upon us the necessity for examining the traditional approaches to disarmament and arms control in this special context.

Legal proscription of ecocide

I am convinced that what is urgently required at this time is the establishment of the concept that widespread and serious ecological debilitation —so-called ecocide—cannot be condoned. I would first limit it to ecocide caused by military activities, although it might well be argued that such a limitation is shortsighted. On the other hand, I would not restrict the concept to intentional military ecocide. Intent may be not only impossible to establish without admission but, I believe, it is essentially irrelevant.

Professor R. A. Falk provides us with a proposed Convention on the Crime of Ecocide ('Environmental Warfare and Ecocide—Facts, Appraisal, and Proposals', *Bulletin of Peace Proposals*, Vol. 4, No. 1, 1973) which would serve to establish ecocide as a crime in international law. A convention of this sort should be designed to complement the Genocide Convention of 1949. Indeed, in some instances it would be difficult to decide which is the more appropriate convention to invoke, as for example in the case of widespread chemical destruction of the agricultural lands of entire primitive cultural groups.

What is next required is legal instruments whose focus is the category of military activity and/or the nature of the target. It appears to me that one of the most useful approaches to preventing future military ecocide is to focus on the target rather than, or in addition to, particular weapons and techniques. Thus I would urge a proscription of any weapon or technique on the basis of whether it devastates a wide area. It is a strategy of area denial— no matter what the method employed—that is likely to be a particular threat to the regional ecology. Moreover, it seems to me that such military activity can be rather clearly and simply defined (an important consideration in the drafting of an appropriate legal instrument) and infractions readily recognized. It is also useful to add here that area denial techniques cannot discriminate well between military and non-military targets or between combatants and non-combatants—further compelling reasons for their prohibition. Indeed, on the basis of the Indochina experience, the major brunt of any environmental warfare is far more likely to be absorbed by the civilian sector of the recipient nation than the military sector.

Once the formidable problems associated with the formulation of legal instruments aimed at preventing ecological warfare are overcome, the task would still remain of having them accepted by the major governments of the world—unilaterally, bilaterally and multilaterally. The problem is that when a government reaches the point of pursuing its aims through warfare, the tendency is for it to employ the most powerful and tactically efficient weapons at its disposal. One crucial restraint on such a tendency, it seems to me, is provided by public opinion; weapons generally considered to be abhorrent are less apt to be employed. A realization of the close interdependence of man and nature is only just beginning to emerge as a result of the increasing stress man is imposing on this relationship. Thus, in the last analysis, I believe that only by arousing world opinion to recognize the abhorrence of any major ecological destruction will we be able to achieve international acceptance of a proscription of ecocide.

4. *Prohibition of weapons of mass destruction**

In a speech made in Moscow on 13 June 1975, the General Secretary of the Central Committee of the Communist Party of the USSR drew attention to a 'serious danger that still more frightful weapons than even nuclear ones may be developed'. He said: 'Reason and the conscience of mankind dictate the need for raising an insurmountable barrier against the development of such weapons'. Accordingly, the Soviet delegation to the thirtieth session of the UN General Assembly suggested the inclusion of a new item in the agenda of the session, entitled 'Prohibition of the development and manufacture of new types of weapons of mass destruction and of new systems of such weapons' and submitted a draft agreement on such a prohibition.

The key article of the draft (Article I) provided that 'Each State Party to this Agreement undertakes not to develop or manufacture new types of weapons of mass destruction or new systems of such weapons, including those utilizing the lastest achievements of modern science and technology.' The types of weapons and the systems of weapons subject to prohibition would have to be specified as a result of negotiations.

Definition of weapons of mass destruction

The first question which arises in connection with the Soviet proposal is what should be considered a weapon of mass destruction. In arms control parlance, these are weapons capable of a high order of destruction and/or capable of being used in such a manner as to kill large numbers of people. The means of transporting or propelling the weapon, where such means is a separable part of it, is not included in the term. At different times also other characteristics of weapons of mass destruction were given, such as, affecting large areas; directed specifically against civilians or most threatening to civilians; and indiscriminate, unpredictable and uncontrollable as regards the consequences.

The problem of definition came up early in the post-war disarmament debate in the United Nations. On 8 September 1947, the USA submitted

* From *SIPRI Yearbook*, 1976.

a draft resolution according to which weapons of mass destruction included 'atomic explosive weapons, radioactive material weapons, lethal chemical and biological weapons, and any weapons developed in the future which have characteristics comparable in destructive effect to those of the atomic bomb or other weapons mentioned above'. The resolution was adopted by the working committee of the UN Commission for Conventional Armaments and subsequently, on 12 August 1948, by the Commission itself, with the USSR voting against. The Soviet Union criticized the US definition as too restrictive and referred to conventional bombs and rockets used in World War II as weapons with mass destructive effects. The Ukraine defined weapons of mass destruction as weapons directed primarily against peaceful populations. The UK suggested that weapons of mass destruction should include only atomic, chemical and biological weapons. A controversy arose with regard to chemical and biological weapons. The USA insisted on making a distinction between deadly weapons and those which were not deadly, such as tear gas or smoke screen, while Australia preferred that the requirement of lethality should be removed from the definition. Since that time, no attempt had been made to evolve a generally acceptable formula. The term 'weapons of mass destruction' was employed in the Outer Space Treaty and in the Sea-Bed Treaty, without its meaning being clarified in the text of the treaty; but it was then generally understood to cover chemical and biological weapons in addition to nuclear weapons which were specifically mentioned. If, however, weapons of mass destruction other than nuclear, chemical or biological were to be prohibited, as suggested by the USSR, a comprehensive definition would be indispensable.

Asked to identify what specific types of weapon the USSR had in mind, the Soviet representative to the United Nations quoted Western newspaper reports about the possibility to develop explosive devices 'from an element even heavier than uranium', or to isolate 'for instance, protons, neutrons or quarks' to produce even more destructive new weapons. But he admitted that some press speculations about new weapons rested on purely fantastic assumptions.

Definition of new weapon systems

In addition to a ban on the development and manufacture of new weapons of mass destruction, the agreement would prohibit new systems for existing weapons of mass destruction. Another question therefore arises, namely, what should be considered a weapon system. In military literature, it is described as a system comprising a weapon, a means of locating and identifying a target, a means of delivering the weapon to the target and means of controlling both the engagement as a whole and at least a part of the sequence of operations that brings the weapon to the target. Furthermore, these components must be so ordered and interrelated as to form a distinct and substantially autonomous system. And, finally, the application of the term is restricted to systems whose target-location or weapon-control functions are largely performed by inanimate apparatus. Assuming that the above description were accepted, one would still have to define the meaning of a 'new system': whether all or only certain components of the system would need to be improved to make it new and, thereby, subject to prohibition.

As examples of new weapon systems, the Soviet representative mentioned binary chemical weapons, 'gene engineering' as a biological weapon, environmental modification techniques, and some unspecified systems of strategic armaments, as well as precision-guided bombs, called smart bombs. It will be noted, however, that: binary weapons are already being dealt with in the context of a comprehensive ban on chemical weapons; possible new biological means of warfare would be best discussed within the framework of the Biological Convention which has been in force since March 1975, and which provides for a review of scientific and technological developments relevant to the convention; environmental modification techniques for hostile purposes are expected to be covered by a separate convention (see unquoted sections); new systems of strategic armaments, such as cruise missiles or modern types of nuclear-weapon-carrying submarines and bombers, constitute the topics of the current US-Soviet Strategic Arms Limitation Talks; and a ban on 'smart bombs' would in all likelihood have to be discussed jointly with other measures of conventional disarmament, unless the definition of weapons of mass destruction has been broadened to include conventional explosives. The prevalent impression in the UN General Assembly was that the sponsors of the draft agreement were not quite sure as to what, specifically, they intended to ban, unless they had decided not to reveal the details of their proposal before negotiations had begun.

The UN General Assembly requested the CCD to proceed, with the assistance of 'qualified' government experts, to work out the text of an agreement prohibiting new types of weapons and new systems of weapons of mass destruction.

The idea of preventing the appearance of ever new instruments of war appeals to many nations, even if these instruments are no more frightful than nuclear weapons.

It would seem that a ban on weapons which have not yet been invented should not be given priority over the prohibition of those in existence, especially nuclear weapons, the mass destructive power of which is subject to no doubt. In any event, a substantial reduction of existing nuclear arsenals would provide stronger incentives for controlling new weapon technology than those which exist today.

5. Conventional weapons and arms trade

1. *The automated battlefield**

Automated warfare is an extremely complex field, and the development of the automatic systems involved draws on advances in many branches of science and technology, such as electronics, telecommunications, computers, chemistry and so on. Automated warfare is also a very dynamic field. In all branches of military technology, the development of a new offensive weapon system is followed by the development of a new defensive system effective against it, and this is further followed by the development of yet another offensive system designed to overcome the defence.

Of the many fields of warfare in which automation has been or could be applied, we will deal mainly with only one—the automated ground combat systems that have been developed, essentially during the past decade.

In order to examine the components of an automated battlefield, it is convenient to consider a battle sequence as being made up of four phases. The first is the location and identification of the enemy, the intelligence-gathering phase. Once the enemy is located, a decision is made on the appropriate course of action to be taken, the decision-making phase. Third comes the action phase, in which weapons are actually fired against the enemy targets. And finally, it is necessary to assess the results of the action against the targets in order to decide whether or not the sequence should be repeated. The last phase is called impact monitoring.

The intelligence-gathering system: sensors

The identification and location of enemy targets on the automated battlefield is achieved by means of artificial sensing devices which will here be called sensors.

The range of types of sensor which can be used in a battlefield situation is very wide. At one end of the scale, there are devices which are really little more than sophisticated 'trip wires': they can be used, for example, in front of defence positions or by an individual patrol camping at night to give advance warning of the presence or approach of an enemy.

Of much more interest in the context of the automated battlefield are sensors that can be remotely deployed and that can transmit information

* From *SIPRI Yearbook*, 1974.

concerning enemy locations or movements over long distances by radio. Such sensors can be designed to respond to a wide range of physical stimuli originating from enemy troops or vehicles, such as sound, seismic disturbances, radio frequency waves, infrared radiation, visible light, chemicals, magnetic fields and so on. The sensors can be either passive, in which case they simply detect signals originating from some external source (for example, a microphone picking up sound), or active devices which themselves emit a signal and then monitor interferences or reflection of this signal due to some external object (the classic example being a radar).

Sensors for use on the automated battlefield should have the following characteristics: they should be as small or as easily camouflaged as possible; they should be able to detect signals from as wide an area as possible; they should be able to transmit their collected information over as long a distance as possible; they should have as long an operating lifetime as possible; they should be able to discriminate between signals originating from different sources; and they should be as cheap as possible.

Readout, processing and display equipment

Sensor systems involving great numbers of sensors reporting information from the automated battlefield require special receivers, capable of monitoring many separate input channels simultaneously, if optimal use is to be made of the information. Each channel in such a receiver accommodates transmissions from a certain number of sensors, each of which identifies itself by means of its own special code. An example of such a receiver is the Portatale: an early model could monitor about 800 sensors and a later generation is capable of monitoring some 40 000.

Decision making

Data from sensors, automatically collected, transmitted, processed, analysed and displayed, forms the basis on which decisions concerning action against enemy targets are taken. Human operators, sitting in control stations remote from or flying over the battlefield, could of course perform this task, and this is the most common procedure at the present time. But modern technology introduces the potential for this phase of the automated battlefield sequence also to be fully automated. A computer may be programmed to react to certain information by issuing orders to other military units. When the computer identifies a target signature, it can transmit information on the location of the target, in terms of grid coordinates, either to a reconnaissance sortie for further investigation, or to a ground unit, a ship, an aircraft, a missile or a remotely piloted vehicle for immediate attack.

This phase of the automated battlefield is the one against which most criticism has been directed. It has been argued that deficiencies in sensor discriminating ability and imperfections in transmission equipment and computers are likely to lead the decision maker, whether a man or a computer, to make serious mistakes in determining counteractions, thereby using weapons indiscriminately.

Counteraction: weapon platforms, weapons and munitions

In order to solve the problems of great distances over which weapons must be fired against targets on the automated battlefield—the major effort has been directed towards utilizing airborne weapon-delivery platforms. These can be either manned aircraft or unmanned remotely piloted vehicles.

F4 Phantom aircraft were commonly used for air attacks against targets identified by sensors in South East Asia. A computer on board the Phantom automatically guided the aircraft to the designated point, using target grid-coordinate data received from a ground control station, and automatically released the weapon load. Sometimes B-52 bombers were also used, in much the same way, because of their great weapon load capability.

Depending on anti-aircraft defences, low speed aircraft have also been used in attack roles. Transport aircraft such as the C-130, and helicopers such as the AH-1G Huey Cobra and the UH-1M Iroquois, have been converted to combine detection and destruction capabilities in single, self-contained night attack systems. These aircraft, often called gunships, are equipped with various sensors and other night-observation devices such as radar, low light level television cameras and forward-looking infrared devices, which enable them to navigate, detect and engage targets in darkness and low visibility conditions. They are armed with multi-barrel guns such as the Vulcan 20 mm aircraft gun—a weapon with six barrels, capable of high rates of fire (a maximum of about 6 600 rounds per minute) with a muzzle velocity of 1 036 metres per second. A further development is the 'Pave Spectre' project, an advanced version of the AC-130 gunship, armed with 40 mm guns and equipped with an on-board computer which can automatically fly the aircraft and aim and shoot its cannon.

A remotely piloted vehicle, equipped with small television cameras, data transmission links and even missiles, can operate in hostile environments with little or no risk to personnel. It can be 'piloted' by operators located in ground control stations or launch aircraft remote from the combat areas. But apart from relieving pilots from exposure to modern air defences, RPVs offer many other advantages. They can be considerably smaller, and in many cases simpler, than manned aircraft, because they do not need to carry such equipment as life support and ejection systems. They are considerably cheaper: it is assumed that a remotely piloted, multi-mission, low-altitude strike-reconnaissance vehicle will cost about $300 000 as compared with more than $15 million for a modern air-superiority fighter. And they can be designed to withstand turning rates and g-forces far greater than those which could be endured by a human. On the other hand, they can be as versatile as a manned aircraft. Tests have shown that an operator in a ground control station, controlling an RPV in real time with the aid of television presentations returned from the aircraft, can guide the vehicle as if he were present in the cockpit; can successfully deliver Maverick and Shrike missiles against targets identified by sensors carried in the vehicle, and can subsequently recover the vehicle. Of course there are still limitations to these systems, such as the restricted field of view of the television camera as compared with a human eye. But the field of remotely piloted vehicles has considerable potential for further development, and is likely to add new dimensions to automated warfare.

Guided weapons

The development of guided weapons—surface-to-surface missiles, air-to-surface missiles and bombs—provides a most effective means for ensuring that a weapon is delivered to its target on the automated battlefield with a very high degree of accuracy.

Certain missiles can be equipped with on-board automatic homing devices and, when launched, these missiles will find their own way to the target without further assistance. Homing systems may be infrared-seeking devices which will guide the missile towards heat sources (in much the same way as the Soviet SA-7 and the US Redeye anti-aircraft missiles are attracted towards the hot exhausts of aircraft engines); radar-seeking devices, as contained in the US AGM-45 A Shrike missile which is used to attack ground radar stations; or inertial guidance systems, similar to those used for aircraft navigation as described above, which are carried in the US Lance Battlefield Support Missile. Different systems can be combined to give better reliability.

Guidance systems of other missiles or bombs must be 'locked in' on a target by an observer, such as the pilot of an aircraft, and this technique offers even greater accuracy in hitting targets. Two major systems have been developed so far for this purpose. The US Paveway series is a family of laser-guided bombs. The target is illuminated with a laser beam from an aircraft, or even from the ground; a laser-seeking device in the bomb or missile then locks on to the laser beam reflected from the target, and a guidance system guides the projectile along the reflected beam to the target. In the second system, bombs, such as the US AGM-62A Walleye, or air-to-surface missiles, such as the US AGM-65A Maverick, are equipped with television cameras. An image of the target as seen by this television camera is displayed on a monitor screen in the aircraft and, using this image, the pilot is able to lock the missile's or bomb's electro-optical tracker on to the target. The accuracies of both of these systems are reportedly very high, with 'circular error probabilities' of no more than a few feet. Both systems also provide the delivering aircraft with stand-off capability. One restriction on effectiveness in both cases is that the targets must be visible. But the use of infrared technology for the tracking and guidance system will provide night operation capability and could also give superiority over other systems under certain daytime poor visibility conditions.

Area weapons

If a target's position is not known accurately, or the target is not visible to an aircraft or the aircraft's navigation system is not capable of determining its *precise* position, highly accurate guided weapons are of less use. The military requirement in these cases is for a weapon which can cause an impact over a rather wide area so that there will be at least some chance of hitting the target. In some cases, guided weapons can deliver such large warheads that they can be considered almost as area weapons. But another means of increasing the area of coverage of a warhead is to divide it into a large number of small 'packages' (bomblets) and to spread these over the area before activating them. This technique was developed during World War II for attacks with incendiary bombs: large numbers of two-kilogramme thermite sticks were distributed over a wide area by a 'mother bomb'. One such bomb is the US M-36 incendiary bomb, containing 182 incendiary bomblets, which was used in South East Asia.

A very large number of cluster bombs, containing fragmentation or hollow charge warheads, incendiary agents, chemical agents and so on have been developed: the US inventory includes more than 80 types in the Cluster Bomb Unit (CBU) series. The CBU-24 bomb (often called the 'Guava' bomb because the bomblets resemble a guava fruit in size and shape) is one of the most widely used antipersonnel and anti-truck munitions in South East Asia. The 600 bomblets contained in this cluster bomb are spread over an area approximately one kilometre long by 300 metres broad, and each bomblet detonates on impact ejecting about 300 steel pellets at high velocity. These pellets puncture tyres, fuel tanks and radiators on trucks and kill or wound the drivers and other persons in the area. Another version has also been developed, with time delay fuses designed to hinder rescue activities. Other weapons include rockets and artillery shells which contain thousands of small steel darts called flechettes.

The above are only a few examples of modern weapon developments which, either because of their high degree of accuracy or their ability to cover wide areas, are not only compatible with the concept of the automated battlefield, but indeed make such a concept more feasible.

Impact monitoring

After weapons have been launched against targets detected and located on the automated battlefield, it is important to know how effective this action has been; if the targets have not been destroyed, then another 'round' of the sequence, beginning again with the gathering of new intelligence, may be necessary. This final phase of the automated battlefield sequence—impact monitoring—could of course be carried out by manned aerial reconnaissance or ground patrols, but on the automated battlefield where the object is to keep personnel remote from the action, it is more logical to use remote sensing devices and data transmission systems, that is, to use the same intelligence-gathering systems that provided the initial target identification and location.

Data collected by these sensors is transmitted back to control stations for assessment in the same way as the initial target detection data. Computers in the control stations can be programmed to compare this data with previous reports and also to initiate the appropriate action, for example, to give instructions to cease fire.

Indiscriminate use of weapons

Automated battlefield techniques used against technologically less sophisticated adversaries will certainly involve fewer military problems than if used against technologically advanced countries. While some of the difficulties that will arise in the latter case will be countered by technological improvements and developments, others, such as the problem of gaining air supremacy and the use of radio frequency waves in an overcrowded radio environment, may remain serious military obstacles for a long time. The increased use of sophisticated sensors is expensive. And counter-measures against sensors and other equipment must always be expected.

A fully automated battlefield, in the sense that human beings are replaced by machines in all phases of the combat, will probably only be feasible in certain restricted areas of the world. Selected techniques and functions, rather than a fully integrated automatic system, are likely to be employed as complements to other combat systems.

The main concern in the recent public debate has been focused on the possibility that indiscriminate use of weapons might follow from the introduction of automated battlefield systems. Even with improved sensor discriminating capability and very sophisticated computers, the confidence in the infallibility of these modern technological achievements is not so well founded that one could dare believe that weapons launched by these systems would not involve the risk of indiscriminate use. Given the long ranges and immense destructive power of many modern means of warfare, any mistake in target designation or weapon launching may have grave consequences. In such cases, when technology is used extensively as a substitute for human senses, all efforts should be made to avoid applications that are likely to promote such risks.

2. *Precision warfare**

Nearly half a million scientists and engineers—roughly half the world's total scientific and technical manpower—are employed on improving existing weaponry and developing new weapons. An effort of this magnitude cannot fail to produce striking results. In the US, for example, the return between 1966 and 1971 (not an atypical period) on military R&D investment was the capability to deploy 11 strategic systems, 14 tactical aircraft systems, 8 tactical missiles and 18 ordnance systems—an average of about 10 new major weapon systems deployed per year. A similar list could almost certainly be compiled for the equally active Soviet Union.

* Shortened version of a paper by Frank Barnaby from *New Scientist*, Vol. 66, No. 948, 8 May, 1975.

New weapon systems emerge not because of a specific military requirement but because of the sheer momentum of the process. For these reasons, the implications of major new weapons for the conduct of warfare, or more seriously, for world or regional security are usually not appreciated until long after the weapons are in the arsenals. This state of affairs is dramatically illustrated by recent advances in a particularly dynamic field of military technology—the development of precision guided munitions (PGMs). These weapons could revolutionise future warfare, particularly if they are used in combination with automated weapon systems.

The development of PGMs was greatly stimulated by the very effective use made of laser-guided bombs in Vietnam. There now exist bombs, missiles, and other projectiles that can score direct hits on their targets with a high probability. According to the design of the PGM, the target may be a tank, ship, aircraft, radar installation, or concentration of armour or troops. A PGM may be an air-to-air, air-to-surface, surface-to-air, or surface-to-surface weapon and may be fired from an artillery piece, aircraft, ship, or vehicle, or be launched by an individual soldier. High accuracy is usually achieved by guidance during the mid-course and/or terminal phases of the projectile's trajectory. By this method, PGMs have reduced the circular error probabilities (CEPs) of some tactical weapons to a few feet—a staggering reduction compared with the accuracy of unguided weapons. Various homing devices, based on electro-optical methods, TV, lasers, infrared, wire-guidance or radar homing, have been developed for PGMs.

A variety of new weapons

Perhaps the best known PGMs are the US Paveway family of laser-guided bombs, popularly called 'smart' bombs. As used in Vietnam, these include Mk 82 500-lb bombs and Mk 84 2 000-lb bombs controlled by a laser receiver. The target was illuminated with a laser beam from an aircraft (either a spotter aircraft or the bomb carrier) or from the ground. The laser-seeking device in the bomb locked on to the laser beam reflected from the target and the guidance system guided the bomb along the reflected beam to the target. High accuracies were obtained.

Also first used in combat in Vietnam was the Soviet SA-7 Grail anti-aircraft missile which was again used effectively against low-flying Israeli aircraft by the Syrians and Egyptians during the October war. This infra-red seeking missile can be shoulder-fired by an infantryman or fired from tracked vehicles. It carries a 2.5-kg warhead. The US Redeye and UK Blowpipe missiles are similar.

The Middle East war also demonstrated the increasing vulnerability of tanks to highly mobile anti-tank missiles, particularly the Soviet Sagger and Snapper missiles and the US TOW and Maverick missiles. Snapper and Sagger are wire-guided surface-to-surface anti-tank missiles, powered by solid-propellant rocket motors. Snapper is fired and guided by an operator who can be remote from the launcher: he sights the target through periscope binoculars and uses a joystick to control the missile, keeping it on the line of sight to the target. The missile, about 1 m long, 14 cm in diameter and weighing about 22 kg, travels at a speed of about 350 km/hr. Snagger is similar in operation but smaller than Snapper—only 70 cm in length and 11 kg in weight, with a 2.7-kg warhead. The missile, often mounted in a battery of six on an armoured vehicle, covers its maximum range of 3 km in about 25 seconds.

Both of these Soviet missiles are slow, however, compared with their American counterpart, TOW (tube-launched, optically tracked, wire-guided). TOW can be used either as a surface-to-surface or as an air-to-surface anti-tank guided missile. It is carried on and fired from a vehicle. The missile, about 120 cm long, 15 cm diameter and 18 kg weight, is equipped with two solid-propellant motors. One motor ejects the missile from its launch tube, but burns out before the missile leaves the tube. The second motor ignites after the missile is well clear of the launch position to protect the launcher from exhaust emissions. This motor very rapidly accelerates the missile up to its maximum speed (believed to be about 1 000 km/hr) and then burns out so that the missile leaves no visible trail. The range of the missile is about 3 km.

Maverick is a sophisticated air-to-surface missile for use against tanks, gun positions and other concentrated targets. The missile, normally carried by strike aircraft, is about 250 cm long, 30 cm in diameter and weighs about 200 kg. A Phantom F-4 aircraft, for example, can carry six Mavericks. Each missile carries a 60 kg conical-shaped, deep-penetrating, high-explosive charge. Maverick is guided by a small television camera in its nose.

Radar-homing PGMs are provided with guidance systems which sense the electro-magnetic radiation emitted by enemy radars and home on it to destroy the radar installation. A further development is the inclusion of a second guidance system in radar-homing missiles, such as the use of an infrared seeker to home on the heat from anti-aircraft gun barrels or from the radar's generator. This would improve the efficiency of the weapon even if the enemy turned off his radar to avoid detection by the anti-radar guidance system.

New types of PGMs now being developed also include long-range cruise missiles; laser-guided artillery shells (a 155-mm shell is currently being tested); and multiple independently manoeuvring submunitions (MIMS). In MIMS, a cluster bomb will dispense a large number of small warheads, each of which will be provided with a terminal guidance system to attack individual targets. And research is underway to develop high-powered laster attack weapons. If the latter ever are successfully developed, they will probably be the ultimate in PGMs.

Emphasis on defence

Concurrent with the development of PGMs, considerable improvements are being made in many other conventional military technologies. Hundreds of surveillance (including sensors), target acquisition, and information processing systems—ranging from field computers to advanced airborne sensors and night vision equipment—are either in operation or under development. Remotely piloted vehicles are being developed for use as attack aircraft and great improvements are being made in command and control capabilities. In combination these developments imply a vast change in the nature of warfare. Some of the ramifications of this are far-reaching.

For example, if precision guidance were used in low-yield nuclear weapons it could remove the 'firebreak' between tactical and strategic nuclear war, with potentially catastrophic consequences. And what would be the implications of automated warfare? If both sides used a full panoply of automatic weapons, how would victory be determined? Would the victor be the one with the biggest arsenal at the beginning of war? If so, are we moving into an era in which the powers will spend even more on weapons than they do now? Very little thought has been given to these issues. But one of the most interesting possibilities is that the new technologies could persuade some nations to opt for a totally 'defensive deterrence'. This prospect arises from the relative cheapness and simplicity of operation of PGMs.

PGMs have some disadvantages. Some anti-tank missiles, for example, are unusable at night and in poor visibility, their launchers may be too vulnerable, and some use small warheads which are insufficiently powerful to destroy tanks completely. But these deficiencies will almost certainly be soon overcome. Then defence will become much cheaper than offence, so much so that the most sensible (for smaller powers, possibly the essential) policy will be to devote military expenditure almost entirely to defensive weapons so that an aggressor would be deterred by the sheer cost of an attack. There is little doubt that it will soon become enormously expensive to maintain large arsenals of both defensive and offensive weapons. Countries will, therefore, have to choose between a credible 'defensive deterrent' or a very considerable increase in military expenditures.

*3. Incendiary weapons**

The UN report

In October 1972 the Secretary-General's report on *Napalm and other incendiary weapons and all aspects of their possible use* was published. Incendiary agents are defined as substances which affect their targets primarily through the action of heat and flame derived from self-propagating exothermic chemical reactions, particularly combustion reactions. Examples are petroleum-based incendiaries such as napalm compositions; metal incendiaries,

* From *SIPRI Yearbook*, 1973 and *Incendiary Weapons*, SIPRI, 1975.

such as magnesium; pyrotechnic incendiaries, which contain an oxidizing agent; and pyrophoric incendiaries, such as white phosphorus and certain organometallic compounds, which ignite spontaneously in air.

Incendiary weapons include bombs, rockets, shells, grenades, bullets, flamethrowers, land-mines, igniters, and so on. That is, incendiary agents are found throughout the entire range of conventional weaponry, and are frequently used together with, or interchangeably with, other agents (such as chemical, biological, fragmentation or high-explosive agents) projected by the same delivery system (cannon, artillery, aircraft, and so on).

In certain circumstances incendiary weapons when used in large quantities are capable of causing massive destruction to both the rural and urban environment. Such destruction is often unavoidably and even deliberately indiscriminate, and may be particularly to the detriment of the civilian rather than the military component of a society. Incendiary warfare is particularly cruel in its effects, notably because of the long period of recovery required for survivors and the high probability of permanent deformity with consequent emotional disorders.

When used in large quantities against urban areas, incendiary weapons have proved to be among the most powerful means of destruction known. Against battlefield targets, the military attraction, particularly of napalm, is their effect on certain kinds of matériel and their casualty effects on personnel, particularly where the precise location of the target is unknown. The report concludes by drawing attention to the social and economic consequences of incendiary warfare. There is a marked disparity between the abilities of the developed and the developing countries both to inflict and to repair the economic damage that may result from incendiary attack.

The Secretary-General's report points out that development continues towards producing incendiary weapons of still greater destructiveness and there may also follow a proliferation of these weapons throughout an increasing number of states.

> The situation is therefore gradually deteriorating and this underlines the urgent need for international consideration of effective measures of disarmament concerning incendiary weapons.

The report indicates that the effects of incendiary weapons may conflict with customary norms as embodied in international law. The massive spread of fire, as well as certain tactical uses of incendiaries, may have consequences that are essentially indiscriminate.

> When there is a difference between the susceptibility to fire of military and civilian targets, it is commonly to the detriment of the latter.

Burn injuries are not only intensely painful but require exceptional resources for their medical treatment.

> When judged against what is required to put a soldier out of military action, much of the injury caused by incendiary weapons is therefore likely to be superfluous. In terms of damage to the civilian population, incendiaries are particularly cruel in their effects.

Further, incendiary weapons may have a variety of toxic and asphyxiating effects through the action of carbon monoxide, white phosphorus, the depletion of oxygen, and so on. Alarm is also expressed at the potential gravity of damage to the rural environment through the massive deployment of incendiaries against forests and crops.

The report suggests that the use of incendiary weapons may be part of a more general problem—the increasing mobilization of science and technology for war purposes.

> New weapons of increased destructiveness are emerging from the research and development programmed at an increasing rate, alongside which the long upheld principle of the immunity of the non-combatant appears to be receding from the military consciousness. These trends have very grave implications for the world community. It is therefore essential that the principle of restraint in the conduct of military operations, and in the selection and use of weapons, be researched with vigour. Clear lines must be drawn between what is permissible in time of war and what is not permissible.

Long history of battlefield use

Until World War I, incendiary weapons had a long history of battlefield use, both against persons and against inflammable targets, such as transport carts, ships, siege towers, and wooden buidings. These uses were for a time eclipsed by the rise of explosives and long range projectiles. During World War I mechanical flamethrowers were used in short-range ground operation, and incendiary bombs were dropped from aircraft on towns. Between the wars the technique of replacing ground troops with air power was developed for 'pacification' operations in western Asia and parts of north-east Africa, operations of the type in which napalm bombs were later to be used; both ground and air flame weapons were used in the Italian invasion of Ethiopia. The technique of bombing cities was developed further in the Spanish Civil War and in the Sino-Japanese War. By the beginning of World War II, the major military powers were equipped with the basic technology and tactics of modern incendiary warfare; during the war, incendiary weapons were used both as weapons of mass destruction and as battlefield support weapons. It was also during World War II that the napalm bomb was developed.

Following World War II, the napalm bomb was used increasingly in air attacks in close support of ground troops, and in air strikes independent of ground operations in rear areas. Very large quantities of napalm were used by US and allied forces in Korea and Indo-China and by French forces in Indo-China and Algeria. Napalm bombs became a standard weapon for combating movements of armed resistance to colonial or central government. In the pacification operations napalm was used in air attacks in an effort both to substitute for as well as to support ground troops. Improvised incendiary devices, such as 'Molotov cocktails', have been used not only by large armies but also by small groups of guerilla fighters and individuals.

Several attempts have been made to restrict or prohibit the use of incendiary weapons by international agreement; indeed the disarmament conference of the League of Nations seemed close to success in 1933. But just as the awesome use of gas during World War I overshadowed incendiary weapons in the post-war legal discussions, so did the atom bomb overshadow them in the legal negotiations after World War II. Chemical and nuclear weapons continue to receive priority in disarmament negotiations.

In the meantime, incendiary weapons—and a great variety of new types of 'conventional' weapons—are being used to kill and maim combatants and civilians. As a result there is a renewed effort to bring incendiary weapons, along with other conventional weapons which may be inhumane or indiscriminate, to the forefront of international negotiations in an attempt to restrict or prevent their further use on humanitarian grounds.

4. *Arms trade with the Third World, 1945–1970**

The arms trade with third world countries is one strand in the multiple set of relationships which connect rich and poor countries, and more powerful and less powerful countries.

The role of the arms trade has been determined by three factors. The first is the increasing complexity of weapon systems. As the cost of developing and producing weapons has increased, production has become concentrated in a few rich countries. Of the supply of major weapons (i.e. naval vessels, aircraft, missiles and armoured fighting vehicles) to third world countries during the period 1950–69, an average of 87 per cent came from four countries—the United States, the Soviet Union, Britain and France—and during 1965–69 this proportion increased to over 90 per cent. Developing countries have had little success in establishing independent armaments bases.

Secondly, the supply of weapons is largely controlled by governments. All military exports require government permission. Not only do governments generally control the defence industry, either through ownership or by virtue of their monopsonistic position, but they also bear a large share of the responsibility for organizing and promoting the export of weapons. Less than 5 per cent of the arms trade is in the hands of private dealers, and only a minute proportion of these dealers operate without government approval.

Thirdly, the two super powers, the United States and the Soviet Union, mostly supply weapons free of charge or at subsidised prices and low interest rates.

The consequences of these three features is this: third world nations wishing to buy or even produce weapons more sophisticated than rifles,

* From *The Arms Trade with the Third World*, SIPRI, 1971.

machine-guns or mortars are dependent on the good will of the governments of a few industrialized countries. In particular, those countries with scarce foreign exchange resources are dependent on the good will of the two super powers.

Main trends

The increase in the flow of major weapons in the last fifteen to twenty years has been substantial: in volume, it was some three times greater in the second half of the 1960's than in the first half of the 1950's. The trend rate of growth from 1950 to 1969 has been about 10 per cent a year with Viet-Nam included, and about 9 per cent a year excluding Viet-Nam.

Behind this rising trend lies an increase in the sophistication of the weapons supplied to third world countries. The term *sophisticated weapons* is used to refer to weapons which incorporate a relatively advanced technology. Technical advances have involved an enormous increase in the cost of manufacturing weapons, as well as an increase in the military efficiency of weapons, in that fewer units may be required to fulfil a particular military role. Thus, for example, there has been a levelling off of the crude numbers of combat aircraft supplied, but at the same time there has been a steady rise in their value—combat aircraft account for a major share of the total flow.

At the same time, there has also been an increase in the quantity of less expensive major weapons supplied to third world countries. The increase in guerilla warfare has led to a corresponding increase in the supply of appropriate weapons—for example, armed trainers, patrol boats, helicopters and hovercraft. These are called *counterinsurgency (COIN) weapons*. Because they are less expensive, and because many COIN weapons are small arms, they do not make a significant impact on the total flow of major weapons.

The main suppliers

Over the past twenty years, the United states has been the dominant supplier of major weapons to third world countries, followed by the Soviet Union. These two countries accounted for about two-thirds of total deliveries. Britain comes next, and then France: these two countries have been responsible for over 20 per cent of total deliveries.

The pattern of arms supplies has changed considerably over the last twenty years. The most important change has been the declining share of the United States in the market for major weapons, and the increase in the share of the Soviet Union. In the first half of the fifties, the United States alone was supplying half the total: the Soviet Union, at that time, did not supply arms to countries outside the socialist camp. In the second half of the fifties, the United States still accounted for 40 per cent of the total. From then on, the US share in the total fell, and the actual volume of US supplies, excluding Viet-Nam, was only marginally higher in the second half of the sixties than in the second half of the fifties. The main reason for this was the increased emphasis on guerilla warfare. There was an increase in the quantity of COIN weapons supplied and a levelling off of the supply of sophisticated weapons. There was also a big increase in supplies to countries involved in guerilla warfare, such as Thailand, Laos and Viet-Nam. The supply of sophisticated weapons may have been somewhat discouraged by the change in the terms under which weapons were supplied to developing countries during the sixties. During the fifties, almost all weapons supplied by the United States to developing countries were free of charge. During the sixties, there was an increased emphasis on the sale of weapons. By 1969, the United States was selling for cash or credit over $800 million worth of arms to third world countries.

Until 1970, the Soviet Union's major weapon supplies were rising very fast. The main increase occurred after 1955, when the Soviet Union began to supply arms to the Middle East. Since then, the number of third world countries which have received major weapons from the Soviet Union has reached a total of 29. By the second half of the sixties, the Soviet Union was supplying more major weapons to third world countries than was the United States. The peak came in 1967, when the Soviet Union replaced the very heavy Arab losses of the June War. Over a third of the Soviet major weapon supplies have gone to the Middle East.

Britain was still an important supplier of major weapons in the 1950's: it was a much less important supplier in the sixties. The United States and the Soviet Union replaced Britain as the main supplier to many of its traditional markets in the previously dependent territories in Asia and the Middle East.

France, on the other hand, although an ex-colonial power, has increased its share of the market throughout the period. From an average of $17 million during 1950–54, major arms supplies rose to an average of $74 million in the second half of the sixties. France has taken advantage of opportunities created by reluctance on the part of other supplying countries. Thus a large proportion of arms supplies went to South Africa after the British embargo of 1964. France made its entry into the Latin American market when the United States was discouraging Latin American purchases of sophisticated weapons, and sold weapons to Pakistan after the US embargo of 1965.

Since the mid-fifties, there has been an increasing number of suppliers as more developed countries, for example West Germany, Italy and Japan, have established arms industries.

Recipient areas

Seventy per cent of all major weapon supplies have gone to the Middle East and Asia. The Middle East has accounted for nearly 25 per cent of total major weapon supplies since 1950, and the Far East, including Viet-Nam, accounted for a further 30 per cent. The Indian Sub-continent has absorbed around 15 per cent of the total.

Major weapon imports to both the Middle East and Viet-Nam have shown an enormous increase. In the Middle East, the yearly trend rate of growth since 1950 has been 15 per cent. The most startling increases occurred in 1955 and 1956, and in 1967. On both occasions, major weapon imports doubled. In recent years, the increase in major weapon imports has been at least as rapid in the oil-rich countries of the Persian Gulf as in those countries directly involved in the Arab–Israeli struggle.

In Viet-Nam, major weapon imports in the second half of the 1960's were twenty-four times higher than they were in the second half of the 1950's. In the rest of the Far East, major weapon imports have declined somewhat since the peak year of 1958 when the United States supplied large quantities of major weapons, in particular large quantities of F-86 Sabre fighters, to the 'forward defence areas' in the Far East, i.e., Taiwan, South Korea, the Philippines and Thailand. In the early sixties there was also a very heavy arms build-up in Indonesia, which has fallen off considerably since the coup of 1965.

Major weapon supplies to South Asia also reached a peak in 1958. India was then receiving Hunter and Canberra bombers and frigates from Britain, partly in response to US supplies of F-86 Sabre fighters, Martin Canberra bombers and destroyers to Pakistan—another forward defence area. Supplies to both countries rose again from 1966. The initial Indian build-up after the 1962 conflict with China mainly consisted of smaller, less expensive items for Himalayan warfare, and these do not all appear in the figures.

Africa, as a whole, has accounted for only 8 per cent of total arms supplies, although this share has been increasing. There has been an arms race in North Africa between Morocco and Algeria; in addition, Libya is now making large purchases of arms. Major weapon supplies to South Africa have increased from $1 million in 1960—the year when many Black African countries achieved independence and the year of the Sharpeville massacres—to $41 million in 1969. South Africa imported more major weapons in 1969 than the total to the rest of Sub-Saharan Africa.

The level of major arms supplies to Latin America was hardly higher in the years 1965–69 than in the years 1950–54. The trends in arms supplies were different in Central America and in South America. In South America, major weapon supplies reached a peak in the second half of the fifties when several countries acquired expensive naval vessels. They have begun to rise again recently as these countries re-equip their air forces and navies. In Central America, major arms supplies reached a peak in 1961–62, when there was a heavy arms build-up in Cuba.

5. *Arms trade policies: supply and demand*[*]

Factors influencing the supply of weapons

There are a number of different policies which govern the supply of weapons from different countries. Each policy represents a varying mix of three main patterns of supply: a mix which reflects the position of the supplying country in the international system.

The first pattern of supply is here called *hegemonic*. Described in abstract terms, a hegemonic system is an international structure in which some countries are relatively dominant and others dependent. It also covers the case where a larger country influences or seeks to influence a smaller country in minor matters: it includes the concept of influences as well.

Secondly, there is an *industrial* pattern of supply. Where it is important to maintain an advanced domestic defence industry and where this is made possible through the export of weapons, then, if this is the *only* function of the arms trade, arms are supplied indiscriminately to any recipient which can afford to pay for them.

Finally, there is a *restrictive* pattern of supply. The characteristic feature of this pattern is that arms are not supplied to countries where this may, directly or indirectly, involve the supplier in a local or international conflict. In contrast to the industrial pattern, it is a pattern of restraint and may often operate by restraining an industrial or hegemonic pattern.

The policies of the United States and the Soviet Union represent a hegemonic pattern of supply. This stems from their leading positions in the two major systems. Both countries can maintain very advanced defence industries without reliance on exports. Their mutual competition, which is extended to third world conflicts, undermines the possibilities for restrictiveness in their arms supply policies. Nevertheless, in the United States there are strong internal pressures for a more restrictive policy—indeed, there may be such pressures in the Soviet Union, although outside observers are in no position to judge.

The arms supply policies of Britain and France are dominated by the industrial pattern. For such medium-range countries, an advanced domestic armaments base is considered necessary to ensure their independence both within and outside the major systems. To maintain such a base the export of arms often plays an essential role. There are elements of hegemony in the international position of these countries, especially with respect to ex-colonies. But the needs of the defence industry, particularly for Britain, have often impaired the function of the arms trade in maintaining hegemony.

Sweden and Switzerland have restrictive patterns of behaviour; their policies are based on the principle that no arms should be supplied to countries in conflict. This pattern arises from their position outside the East-West struggle, which could be compromised by involvement in local conflict.

---

[*] From *The Arms Trade with the Third World*, SIPRI, 1971.

West Germany, Japan, Italy and Canada have conflicting industrial and restrictive patterns of supply. In practice, the policies of West Germany and Japan tend to be the more restrictive; involvement in local conflicts could harm their commercial and political interests in the third world, especially since these two countries have a historical image as 'warmongers'. In addition, their industrial requirements can largely be met by exports to their Western allies. The Italian and Canadian restrictive policies are selective, both with respect to certain types of equipment and certain areas.

There are additional benefits to be derived from supplying arms which may not be embraced by these three patterns of supply. One is the desire to get better than scrap value for surplus equipment. Another is to increase foreign exchange earnings. It may also be useful to gain field experience of weapons.

Hegemony

There are three ways in which the supply of weapons can reinforce hegemony. First, arms can be provided to enable local forces to perform military tasks which are in the interests of the supplying country: to prevent real or imagined military threats to the system or to expand the system by force. Secondly, the supply of arms may serve to strengthen the relationship between the supplying country and the recipient government. In this case, the supplying country may have little or nothing to gain through the use of weapons. Having received arms, the recipient country becomes, in some degree, militarily dependent upon the supplying country, which can then demand favours or threaten to withhold spares or further supplies if the recipient does not comply with its interests. Alternatively, the supplying country may simply be interested in preventing another supplier from achieving this kind of relationship: this is known as pre-emptive supply. Thirdly, the supply of arms may provide an opportunity for influencing individuals in the recipient countries, in particular those in the military establishment. This occurs when contacts are made between the military personnel of the two countries and when military training accompanies the provision of weapons. This function is most relevant for countries where the military play an important role in politics, as they do in a large number of countries. Inevitably, all these functions of the arms trade overlap.

The United States and the Soviet Union

From the initiation of the US military aid programme, emphasis has been placed on equipping local forces to perform military tasks in the US interest, on 'substituting for what in many cases might otherwise be a vastly more expensive direct American military presence'. Between 70 and 80 per cent of US exports of major weapons have gone to the countries bordering on the Soviet Union and China, known in the USA as the 'forward defence areas'.

Initially, Soviet military aid was also provided to countries where the weapons could be used in the interest of the Soviet Union. But increasingly, this became a limited function because it was confined to the socialist developing countries.

One of the main constraints against this function of military aid appears to have been the Soviet desire to avoid direct confrontation with the United States. Military aid to North Viet-Nam was restrained in the early sixties, as was Soviet support for national liberation movements.

The notion that military aid is provided to enable local troops to perform functions that are in the interest of the supplying country depends on the assumption that either the interests of the recipients are identical with those of the supplying country or that the supplying country is in a position to control the forces in recipient countries. This was not the case with all the recipients of US military aid in the forward defence areas. Many members of SEATO and the Baghdad Pact, now CENTO, were more concerned with their non-communist neighbours. Arms were supplied in order to secure the allegiance of recipient regimes rather than to secure objectives dependent on their allegiance. Similarly, as the Sino–Soviet split widened, the Soviet Union could no longer influence the use of force by North Viet-Nam and the liberation movements through restraining arms supplies. Soviet arms supplies to these recipients increased in order to prevent them from moving into the Chinese sphere of influence—to secure their continued allegiance.

The use of the arms trade as a means to secure the allegiance of a recipient regime depends on the degree of monopoly maintained by the supplying country. As competition between the United States and the Soviet Union intensified, this function increasingly degenerated into pre-emptive supply, that is, the use of the arms trade to prevent a country from becoming the client of another country.

The industrial pattern

The importance of exports to the domestic defence industry has been stressed in all European countries. This partly arises from the substantial economies of scale in the production of weapons. Domestic demand is generally insufficient to allow the production of advanced weapon systems without exorbitant cost or without confinement to a narrow range of weapons. Over the last two decades, the costs of developing and producing weapons have steadily increased, while the number of units required to carry out a given military task has declined.

Exports lead to a lengthening of production runs. There are two main benefits derived from long production runs. First of all, the costs of Research and Development and the special installations required for the production of particular types, which represent over 30 per cent of the total costs of producing and developing a typical military aircraft, can be spread over a larger number of units. Secondly, productive efficiency increases as more units of a particular product are produced; that is, labour and material costs per unit fall as experience is gained.

Exports also have another important function in maintaining a domestic industry: they can ensure full employment of capital and labour and can protect the industry from fluctuations in government demand. This last factor is possibly more important than the benefits of economies of scale.

Factors influencing the demand for weapons

With the big exception of Latin America, most of the developing world was under colonial rule before and during World War II. The multiplication of nations led to the multiplication of military units under independent control. Moreover, the circumstances in which these nations were created often encouraged the establishment or expansion of armed forces. For some countries, the acquisition of weapons was limited by the scarcity of resources. In others, the competition between the major powers facilitated the acquisition of weapons because of their readiness to provide military aid.

Inevitably, in attempting to provide a broad picture of the factors influencing the demand for weapons, generalizations, not applicable to all countries, are made. Social structures, ethnic divisions and historical circumstances vary enormously. The purpose here is to try to distinguish certain common factors in the requirements for weapons.

These requirements often arise from some form of conflict in its broadest sense. Many of the new nations were created artificially; that is to say, their borders were not necessarily determined by geographical, ethnic or historical considerations, so that internal and external divisions were often built into the nation. In addition, the process of decolonization brought with it abrupt changes in social structure which also caused internal instability.

In such situations, the acquisition of arms can satisfy three sorts of requirements. First of all, there is the purely military requirement; arms are needed for those external and internal conflicts in which force might be used. Secondly, the acquisition of arms can serve to unite divided groups by affirming the national identity. Although national independence movements were often based on mass parties, these parties were united on little save the struggle for independence. After independence, when ethnic and class divisions began to appear, the mass base often began to be whittled away. In bidding to retain their support or to mobilize new support, the new leaders often renewed their appeal to nationalism. Because armed forces represent one of the main attributes of independence, the acquisition of arms is one way of doing this. Related to this is the need for strong armed forces to provide a backing for foreign policy. Thirdly, internal divisions are often reflected in the role of the armed forces. Governments can gain the support of the armed forces by satisfying their demand for weapons.

Independence and hegemony

The importance of nationalism and political independence to certain groups has been stressed. The possession of strong armed forces is considered an attribute of political independence. Yet the acquisition of weapons from outside powers may bring about a state of dependence. Where the interests of the regime in the recipient country and the supplying country coincide, this may not matter. Where they do not, the independence of the recipient country and thus the interests of the regime may be threatened. There is a risk that the supplying country may for political reasons terminate a contract, discontinue the supply of spares, or 'over-charge' for spares deliveries.

Several of the more advanced third world countries have attempted to develop their own domestic defence production: the most important examples are Argentina, Brazil, India, Indonesia, Israel, the UAR and South Africa. The main reason for doing so is to increase military self-sufficiency and reduce dependence on outside suppliers.

There are two ways in which recipient countries attempt to acquire military independence. Domestic defence production tends to defeat its objective not only because it is very expensive, even in terms of foreign exchange, but also because it requires heavy inputs of foreign assistance. Another way is diversification of the sources for weapons. This is limited in two situations. It is limited where the recipient country is dependent on the supplying country in many different fields. In this situation, the supplying country can, if necessary and it usually is not, apply sanctions against a recipient country which turns to other sources. Diversification is also limited where the recipient country is involved in an armed conflict, both because of the military difficulties in switching suppliers and because of the reduction in the number of potential suppliers.

Where hegemony is established through a broad range of transactions, as in Latin America and South East Asia, the specific contribution of the supply of arms to hegemony cannot be isolated from the effects of other transactions. Elsewhere, the continuing contribution of arms supply to hegemony has marginal significance except where the supplying country has a monopoly position, or in so far as the supply of arms prevents the monopoly of another supplying country. Since such monopolies rest on the military situation in these countries, their advantages are offset by the risk attendant on armed conflict. The risk of defeat ties the supplying country to the recipient as much as *vice versa*. It also increases the risk of confrontation between competing supplying countries.

Wars

Weapons are for use in war. Perhaps the most important question about arms supplies is what effect they have on the development of wars—on the likelihood of wars breaking out, on the course of wars and on their general severity. This includes not only the general effect of arms races on wars—which is the same whether weapons are produced domestically or purchased abroad—but also the consequences of supplier entrammelment in conflicts, via the arms trade.

Another important facet of arms supplies is the way in which they draw supplying countries into the conflict. The supply of weapons to countries in conflict can be seen as an indirect use of force. By providing weapons to a particular recipient in conflict, the supplying country implicitly lends support to that recipient. When that support becomes explicit, it may have enormous implications for the course of the conflict.

6. *The arms build-up in the Middle East**

Throughout the post-war period the area generally known as the Middle East has been a focal point of international attention. The region has long been regarded as strategically important both geographically, as the link between Europe, Asia and Africa, and increasingly over time, as a major source of oil supplies. The area was marginally involved in World War II but was directly and continuously involved in the subsequent Cold War with the bitter Arab–Israeli dispute providing a situation which the major powers were long inclined to exploit rather than attempt to resolve.

These circumstances, together with historical regional tensions and rivalries, have produced an exceedingly rapid and enduring expansion of the military forces in the countries of the region. Until quite recently the arms race and repeated wars between the Arab countries and Israel focussed attention on the area. This competition continues with undiminished vigour but it has recently been supplemented by an even more spectacular arms build-up in the countries of the Persian Gulf area, largely as a result of the huge increase in oil revenues which greatly reduced financial restrictions on national ambitions.

Military expenditure

In 1974 (the most recent year for which relevant data are available) per capita gross national product (GNP) for the Middle East region as a whole was about $845 and per capita military expenditure about $135. In other words, nearly 16 per cent of the combined GNPs of the Middle East countries was being devoted to military purposes. This extraordinary degree of militarization is the result of a steep and virtually uninterrupted upward trend in military expenditures over the past 25 years.

On the average, the quantity of financial resources devoted to armaments in the Middle East has increased at an annual rate of nearly 16 per cent since 1950. Between 1950 and 1961 the average annual rate of increase was comparatively modest at 11.5 per cent, although even this rate was nearly double the world average over the same period (6.3 per cent). Since 1961, however, the average annual rate of increase has been 19.5 per cent, or about seven times the world average for the same period (2.8 per cent).

There is no parallel in the post-war period for sustained rates of increase in real military expenditure such as these. To find a comparable rate of

* From *SIPRI Yearbook*, 1976.

escalation in the resources devoted to military uses it is necessary to go back to the arms race that preceded World War II when the major industrial powers tripled their combined military expenditures in the five-year period 1933–38.

Military arsenals: quantitative and qualitative aspects

Quantitatively the flow of weapons into the Middle East has been extremely large, the result of rapidly increasing regional military budgets and intense great-power rivalry yielding large weapon transfers on a grant or concessionary basis. SIPRI arms trade records show that about 4 100 jet combat aircraft (excluding trainer versions of combat types) were transferred to Middle East countries between 1950 and 1975. Similarly, culmulative deliveries of heavy, medium and light tanks amount to about 13 500.

To establish existing force levels of these weapons is considerably more difficult. Over a 25-year period equipment wears out and is destroyed in accidents. Also the Middle East region has experienced three major wars, all of which involved heavy losses of equipment. And finally, of course, the arsenals are not static due to deliveries of new equipment and the transfer of older equipment both within and out of the region. Nevertheless it appears that operational jet combat aircraft in the Middle East numbered about 2 300 at the end of 1975; the equivalent number for tanks was approximately 10 500.

To give these force levels some perspective it can be pointed out that they compare favourably with NATO forces in Europe: approximately 3 000 tactical aircraft and 12 250 main battle tanks.

Apart from their huge size, the other striking feature of the military arsenals in the Middle East is the high proportion of up-to-date equipment despite the fact that no country in the region (with the partial exception of Israel) has any indigenous capacity to develop and manufacture major weapons.

It is clear that, in terms of major conventional weapon systems, the Middle East arsenals are among the most up-to-date in the world. Of the 2 300 front-line combat aircraft mentioned earlier, well over one-half are new-generation MiG-23s and F-5Es and late-model MiG-21s, F-4s, A-4s and Mirage IIIs. Similarly nearly one-half the number of tanks are current types—T-62, M-60, AMX-30 and Chieftain. However, in recent years several Middle East countries have gone beyond the acquisition of advanced major weapons to seek military capabilities usually associated only with the major powers and weapons and equipment that are representative of the current frontiers of military technology. Israel and Iran are in the vanguard of this movement but several other countries in the region are clearly determined to go in the same direction.

Defence industry and military infrastructure

Throughout the post-war period the countries of the Middle East have been utterly dependent on external sources for their armaments. This is still true at the present time and will undoubtedly remain true for the foreseeable future. Nevertheless several countries are determined to reduce this dependence, and to base their military capability to the extent possible on indigenous facilities.

Israel is by far the most advanced in this regard with an indigenous capability to maintain, overhaul and repair virtually all its military equipment. In addition, it has a significant and growing indigenous weapon development and production capability. In Egypt a major expansion in the defence industry is planned through the Arab States Military Industrial Organization with an aircraft assembly and repair plant in Egypt and related facilities in other participating countries, notably Saudi Arabia.

In Iran virtually every major weapon deal concluded in recent years has included the establishment of an appropriate maintenance and repair facility and the training of Iranian personnel in the relevant skills. In the Persian Gulf area, particularly in Iran, huge sums are being spent on army, air force and naval bases, headquarters and command centres and communication networks.

Conclusions

The arms build-up in the Middle East shows every sign of being out of control. The futility of attempts to avoid conflict by trying to preserve a dynamic military balance between opposing parties has been amply demonstrated by the three Arab-Israeli wars. The complexity of the present situation has in any case reduced the feasibility of maintaining some form of balance virtually to zero unless, of course, the arms-supplying countries—particularly the USA, the USSR, France and the UK—all agreed to stop or limit their supplies.

The whole Middle East region is in a state of extreme flux as many of the countries embark on ambitious programmes to modernize and industrialize their societies. Thus, massive structural and societal changes will take place together with very large and growing military arsenals. Both these phenomena increase the probability of regional instability and conflict. Certainly the continued unrestrained sale of armaments can only exacerbate an already dangerous situation.

*7. Proliferation of Vietnam-War related weapon technology**

The Indo-China War led not only to a major investment in new weapon technology but also provided a proving ground to test new weapons. There can be little doubt that this investment in limited war and COIN

* From *SIPRI Yearbook*, 1974.

(counterinsurgency) technology by one of the world's major suppliers of arms will have a significant impact on military doctrines and procurement policies throughout the world.

Two kinds of proliferation are in evidence. The first is the profusion of new varieties of munitions, delivery platforms and target acquisition and guidance systems. Table 5.1 lists some weapon developments which were 'stimulated or accelerated by the pressure of the war in Southeast Asia'. A wide range of other weapons, either developed by government laboratories or by private industry, were tested and employed in Viet-Nam by US and allied forces: examples range from CS gas munitions and herbicides to new kinds of small arms and ammunition, such as the 5.56 mm M-16 rifle and a 12-bore shotgun cartridge containing 20 flechettes.

New Soviet weapons did not appear in Viet-Nam in anything like the same profusion. Weapons like the 'Grail' heat-seeking anti-aircraft missile

Table 5.1. US weapon developments stimulated or accelerated by the war in Indo-China[a]

Tactical aircraft systems
Cobra attack helicopter
A-37 attack aircraft
AC-119/123/130 gunships
OH-6/OH-58 helicopters
Tactical missiles and ordnance
M-16 rifle
M-72 light assault weapon
7.62 miniguns
Standard ARM missile
Talos ARM missile
MK 36 destructor mine
BLU-61 bomb
Rockeye munition
Laser-guided weapons
Walleye/Hobo guided weapons
Fuel-air explosive munitions
CBU 24/49 cluster munitions
Wide Area Anti-Personnel Mines (WAAPM)
20 mm gun pods
2.75-inch rocket with warhead
New bomb fuses
Combat vehicles
M551 Sheridan vehicle
Sensors, Command, Control and Communications, and Intelligence
Loran D navigation system
PPS 4/5, TPS 58 surveillance radars
3 types Starlight Scopes
Laser rangefinder

[a] All these weapons have been adopted by US forces. Some other weapons were tested in Viet-Nam on an experimental basis only.

Source: Statement of John S. Foster, Jr., Director of Defense Research and Engineering, US House of Representatives Committee on Appropriations, *Department of Defense Appropriations for 1972*, Hearings, part 4 (Washington, US Government Printing Office, 1971) pp. 728–30.

and the 'Sagger' wire-guided anti-tank weapon, both light infantry weapons, appeared in Viet-Nam only in 1972. The North Viet-Namese never acquired missiles fully able to defeat the B-52 bomber, which, since it was developed as a strategic nuclear bomber, must have been a major focus of Soviet anti-aircraft-missile development efforts. The new technologies to emerge from the Viet-Nam War were those of the United States.

Many of the new weapon systems developed for the Viet-Nam War by the USA together form what has been termed collectively the 'automated battlefield'.

The second kind of proliferation is the spread of the new weapons and technologies to other countries. A few examples of the spread of the new technologies are the following.

1. Cluster bomb munitions have been developed in France (the Giboulée) and the UK (BL755) and supplied to several other countries. US-made munitions of this type were used by Israel in the recent conflict. The logic of a cluster of small bombs to give greater area coverage is commanding, because greater stand-off capability is demanded for aircraft engaged in attacking small point targets such as men or vehicles.

2. The technique of casting metal balls in a plastic shell, used in some US cluster bombs instead of the traditional fragmentation principle, has been adopted in the West German Diehl-DN-51 hand grenade.

3. The notched steel wire which breaks into small pieces, used in the US M-26 hand grenade and 40 mm grenade, is used in the Belgian PRB 423 hand grenade and in a new Swedish Bofors 40 mm grenade, which also contains metal balls.

4. The widespread use of the 40 mm grenade in Viet-Nam has led to the development of 40 mm grenade launchers for all the 5.56 mm rifles, and a number of larger automatic launchers for aircraft.

5. M-16 rifles, using the very high velocity 5.56 mm ammunition, have been reported in many armed forces, including those of Lebanon, the Philippines, Portugal (in Angola) and the UK (in Aden and Indonesia). A great variety of similar weapons is now being produced by private manufacturers. Israel has adopted its own 5.56 mm calibre rifle, the *Galil*, which was used in the recent conflict.

The end of the overt US involvement in the war in Indo-China may lead to an increased rate of proliferation of many of the new weapons in several ways. First, the decline in contracts for the supply of US forces may stimulate US manufacturers to seek alternative markets. Second, a considerable amount of surplus matériel from Indo-China is being given as military assistance or sold (at one-third cost) to other governments. Third, the USA is likely to focus its attention on other areas, such as Europe, seeking to adapt the new technologies to other theatres. Fourth, other armed forces are likely to seek to acquire the new technologies and other manufacturers to supply them. The processes are in turn likely to stimulate the development and spread of Soviet weapons.

In conclusion, it should be pointed out that light-weight weapons designed for jungle warfare may also be suitable for urban counterinsurgency. Rifles of the US 5.56 mm calibre have been reported in Northern Ireland, and anti-aircraft missiles of the Soviet SA-7 type were discovered in Rome, apparently to be used in an attack on a civil aircraft. The arms race in means of internal warfare is a further example of the proliferation of Indo-China-War-related technology.

*8. Local wars in Asia, Africa and Latin America, 1945–1969**

The present study wishes to deal only with the wars that have taken place in the 25 years since the Second World War on the three former colonial or semi-colonial continents: Asia, Africa and Latin America. Talking about local wars fought on these three continents in the post-war period is, by the way, nearly the same as talking about the local wars of this time on a world-wide scale. Because, though a great many wars have been and are being waged in the world since 1945, Europe has very seldom been a theatre of war in this period of time.

In the period under consideration, it was only in Greece that a significant war of a short duration, at the turn of 1944 and 1945, and a civil war of more than three years, between 1946 and 1948, took place on the European continent. Along with these, a not very extensive but fairly permanent guerilla war flared up again in Spain between 1945 and 1948. Further on, the fights in Hungary in October 1956 came close to, or even reached, the concept and degree of armed conflicts. Other wars, even of a short duration, have not been waged in Europe in this period if we take Cyprus, as we do in our study, to belong to the Middle East.

From the point of view of our study, we have regarded as wars *any armed conflicts in which all of the following criteria co-occur:*

(*a*) The activities of regular armed forces (military or police forces, etc.) at least on one side, that is, the presence and engagement of the armed forces of the government in power.

(*b*) A certain degree of organization and organized fight on both opposing sides even if this organization extends to organized defence only.

(*c*) A certain continuity between the armed clashes, a strategic-tactical coherence between the armed conflicts, however sporadic. In other words, the presence of a planning and organizing central activity on both sides. We have also considered as such the centrally organized guerilla forces insofar as they extended their connected activities over a considerable part of the country concerned.

---

* Shortened version of a study by István Kende published by the Center for Afro-Asian Research of the Hungarian Academy of Sciences, Budapest 1972.

The number of casualties

No attempt has been undertaken to make calculations on the number of casualties, on the quantity of ammunition and equipment used, nor the size of forces engaged since it is only for a few wars that there are relatively reliable data available. Yet, merely by way of estimate, we would risk the statement that the number of persons killed in action in the twenty-five years since the Second World War reaches or, at least, comes close to the total loss in lives in the Second World War. Let us remind that, according to estimates or reports published, the number of casualties of the Chinese Civil War (1946–49) is about 10 million, that of the Korean War (1950–53) 1 to 1.5 or more million. Equally high is the number of killed in the wars in Indo-China and Vietnam, in Nigeria, the Congo, etc. The number of casualties, suffered in other wars as estimated with reasonable accuracy, is as follows: the Indian War of religion (1946–47): over half a million; the Columbian wars: at least 300 000; the wars in Portuguese Africa, Angola, Mocambique, and Guinea (Bissau): several hundreds of thousands. This will suffice, we think, to make our statement acceptable and believable that the number of persons killed in the various wars since the Second World War amounts to tens of millions, and is therefore by no means far from the number of those killed in the Second World War. Thus the totality of local wars corresponds in this respect to a sort of a *Third World War* fought under specific circumstances. Similar conclusions can also be drawn, by the way, from the size of the territories involved.

In the quarter of a century since the Second World War, the territory—or part of the territory—of 59 countries on the three continents became theatres of war. The First World War was fought in 14, the Second World War in 40 countries.

Let us add: though in the time under consideration 93 wars (including Europe and counting thereby the whole world: 97 wars) were fought, *not a single declaration of war was made.*

Duration and distribution by regions

The *total duration of the 93 wars* fought wholly or partly on the three continents since the Second World War up to December 31, 1969 (calculated by adding up their individual lengths) amounts to *255 years and seven months less a few days.* This means that the average duration of a war was *2 years and 9 months.*

This is, of course, only a statistical average. Most of the wars lasted longer than one year: 49 out of 93; less than half of all wars, 44, lasted less than one year.

If we spread the total duration of the 93 wars (i.e. more than 255 and a half years) evenly over the twenty-five years, we would get an average value of *10.22 wars for each day of these twenty-five years.**

The average figure calculated above is a very long time, the average time of wars in a quarter of a century. It is far more informative to point out how this average value—the number of wars per day—changes if counted by periods. For this purpose we have divided the twenty-five years into five-year periods.

Table 5.2. Total duration of wars by five-year periods

Periods	*Total duration by periods*			*Average by year*	*As percentage of total duration*
	year	*month*	*day*		
1945–49	25	7	20	5.13	10.03
1950–54	33	8	18	6.74	13.19
1955–59	48	8	23	9.75	19.07
1960–64	57	8	29	11.55	22.60
1965–69	89	8	24	17.95	35.11
Total	255	6	24	10.22	100.0

Until now, we have treated the three continents as a uniform whole. Now, we must specify our data according to regions.

The four regions are as follows:

—*Asia* (not including the countries of the Middle East),

—*the Middle East* (including all Arab countries, that is both the Maghreb countries and the Sudan and further Iran, Cyprus and Israel),

—*Africa* (excluding the Arab countries, so that what we mean by 'Africa' is the so-called 'Black Africa'),

—*Latin America.*

Table 5.3. Number and total duration of wars by regions

Regions	*Number*	*Percentage*	*Total duration of wars*			*Percentage*
	of wars		*year*	*month*	*day*	
Asia	29	31.2	112	9	3	44.1
Middle East	25	26.9	52	2	28	20.5
Africa	16	17.2	54	–	13	21.1
Latin America	23	24.7	36	6	10	14.3
Total	93	100.0	255	6	24	100.0

* Kende lists 119 wars during the period 1945–75. The total duration of these conflicts exceeded 350 years. The territory of 69 countries and the armed forces of 81 states were involved. Since September 1945 there has not been a single day in which one or several wars were not being fought somewhere in the world. On an 'average day' about 12 wars were fought.

The number and duration of wars are not only very high after the Second World War, but their trend also shows a fairly consistent and steady rise between 1945 and 1969.

Correlation between the dynamics of wars, military expenditure and arms trade

Arms supplies to developing countries whether in the framework of military aid or normal trade cannot be separated from the wars fought on the three continents under examination. Any party participating in or preparing for a war, wishes to get arms for aggression, defence, gaining independence, crushing the forces of liberation or whatever its aim may be. Similar circumstances exercise an impact on other forms of military spending too.

We have taken all military-economic data to be used in our investigation exclusively from one single source—the Yearbook of 1969–70 of the Stockholm International Peace Research Institute (SIPRI).

We have examined the correlation between the trends in wars and the military sectors of economies in two respects:

(*a*) First, we have taken into account the *military expenditure* of the countries of the four regions, supplementing them by data for any *major military aid* received by the same countries; thus we have not separated the two factors, budgets and military aid since the destination of all sums deriving from these sources is the same.

(*b*) Second, we have considered the *arms imports figures* of the countries of the three continents, taking into account the so-called 'major weapons' only, since no information was available on other weapons.

We have found a closer than expected correlation between the dynamics of wars already pointed out and the trends in military expenditure and arms trade. When examining the aggregate figures for the four regions we find a really striking similarity between their trends, especially those in wars and arms imports. But an analysis by regions also corroborates the justification of a comparison of the three trends, even if a more pronounced variation can be found in them in this comparison.

Military expenditure and *military grant aid received* show a very consistent rise, examined both globally for the three continents and also separately for the four regions. The steepness of this rise is, of course different from region to region. It is especially steep for the Middle East in the last years and less steep, though still rising substantially, for Latin America.

Conclusions

We shall confine ourselves to summarizing and partly repeating the lessons and main general conclusions that can be drawn from the data and facts published in our study.

As far as their general trend is concerned, *the number and duration of local wars are strongly and rapidly rising* in our days; there has been some decrease in the past few years, but this has not brought about any appreciable change in the rapidly rising general trend.

The war-type of our time is no longer the 'international' ('frontier') war fought between neighbouring countries; their number and primarily their share in the total duration of wars is relatively insignificant. *The main type of war in our age is the war fought within the boundaries of a single country with the aim to overthrow a given regime or government and bring about changes in power. Predominant are,* within this type, *wars waged with the participation of the forces of foreign countries.*

There is a clear connection between the spread of local wars and military expenditure or military grant aid. This relationship can be especially convincingly shown between the trends in war time and arms imports.

6. Armament dynamics and military research and development

1. *The dynamics of armaments**

Since 1948, the quantity of resources devoted to military uses throughout the world has more than trebled. Along with this increase in world military expenditure in absolute terms, there has been a dramatic increase in the share of world output devoted to military uses. In 1972, world military expenditure was valued at about $207 billion, which represented about 6.5 per cent of total world output. In many of the post-war years, the proportion was even higher—around 7–8 per cent. These are much higher proportions than those which prevailed in any other peacetime period in recent history; in the years preceding World War I, and in the inter-war years, the proportion was some 3–3.5 per cent. The largest increases in military expenditure have occurred in the United States and the Soviet Union. The United States and the Soviet Union account for approximately two-thirds of the total world military expenditure.

Technological pressures

A widely accepted explanation of the trend in armaments in the United States and the Soviet Union is that the 'arms race' in which these two countries are engaged follows a characteristic pattern of action and reaction, or, in most cases, over-reaction. This implies that Soviet military activities are determined almost exclusively by those of the United States, and *vice versa*. The tendency to over-react arises because there is no precise answer to the question of how much (military capability) is enough, and where national security is concerned, uncertainty typically leads to a preference for too much rather than too little.

While the action/reaction (or over-reaction) pattern is undoubtedly a factor influencing the US–Soviet arms race, there is little evidence to suggest that it is the most important one. In the early post-war years, when an atmosphere of fear and mistrust existed in relations between these two countries, the action/reaction process was a major force behind the arms race, but the fact that the level of armaments has not been noticeably influenced by the marked improvements in the international climate in recent

* From *SIPRI Yearbook*, 1974.

years supports the conclusion that it is now of less importance. Instead, it seems that each side, rather than simply trying to match, or improve on, the capabilities of the other side, is now more concerned with exploiting to the full the technological opportunities available to it. This leads to much more rapidly increasing sophistication of weaponry: an action/reaction pattern would, in fact, represent a relatively constrained form of arms racing, since it only requires each side to 'move ahead' of the other by one 'step' at a time; but with each side exploiting technology for its own sake, this constraint is removed. There are a number of factors, such as those described below, which act to keep the technological process going and accelerating, and these are only reinforced, not controlled, by the action/reaction process.

The intense development and exploitation of technology for military purposes is essentially a post-war phenomenon. In the inter-war period, military support for science and technology absorbed less than 1 per cent of the military budgets of the major powers, but the dramatic illustrations of the military potential of science and technology during World War II led to a phenomenal increase in such support. The shares of the major military budgets devoted to research and development (R&D) reached 10–15 per cent by the late 1950's and have remained at this level since, reflecting the widespread faith in the military and political utility of technological sophistication in weaponry. The United States and the Soviet Union have dominated this process.

The observable result of this effort—an unprecedented rate of technological change in the field of armaments—is reflected in the size of military budgets. Thus, the three-fold increase in world military expenditure in the post-war period has largely been due not to increases in the numbers of men under arms or to increases in the quantities of weapons deployed, but to qualitative improvements in weaponry, with each successive 'generation' of weapons costing more to develop, to manufacture, and to operate and maintain.

Long lead-times

The first and perhaps most important characteristic of the technological process that promotes a rising level of armaments is the extremely long lead-times or gestation periods of modern weapons. The development of a new missile or a new fighter aircraft to the stage at which it is ready for mass production and deployment may take up to 10 years. This time factor introduces an important qualitative change in the action/reaction pattern: participants in the 'technological arms race'—primarily the United States and the Soviet Union—focus their attention not on which weapons the other side has already produced or is ready to produce, but on possible future developments in the opponent's weaponry, and then undertake programmes designed to produce weapons to offset these anticipated developments. Thus the uncertainties surrounding the question of how much is enough are increased. Since the range of conceivable developments in the opponent's weaponry is essentially unlimited, there is a *prima facie* case for initiating programmes to protect oneself against as many conceivable developments as possible. The process is only exacerbated by the fact that technological advances make it possible to consider protection against increasingly remote contingencies.

The fallacy in this procedure is, of course, that the other side is following exactly the same pattern. Thus there is a high probability, given comparable technological opportunities and constraints, that both sides will independently pursue similar courses resulting in mutual, and therefore 'justified', advances in weaponry. In fact, the United States and the Soviet Union do have roughly comparable weapons, for this reason. Any imbalances in weapon capabilities that do occur merely supply additional fuel to the process.

The 'follow-on' imperative

Another channel through which emphasis on technology promotes a rising level of armaments is the so-called 'follow-on imperative'. The development and production of modern military equipment is an extremely complex task requiring highly skilled, specialized resources. The technical and industrial teams engaged in these activities are regarded as national assets which, given the dynamic element in the technological process, cannot be allowed to disintegrate without risk to the nation's ability to keep abreast of foreign developments in weapon technology. In other words, these resources must be kept fully employed all the time, and this will automatically lead to the continual development of new weapon systems. An exacerbating phenomenon in this context is the fact that, as weapons become ever more complex, the technical and industrial capacity created at the peak of a programme grows larger. The main reason for this is that, despite the growth in complexity, it is considered impossible in an environment of rapid technological change to make any commensurable increase in the length of the development process: to do so would be to risk the weapon becoming obsolete before it is deployed. Therefore, not only must these specialized resources be kept fully employed, but the increase in their size cannot be controlled.

Internal bureaucratic and economic pressures

After World War II, and to a greater extent after the Korean War, both the United States and the Soviet Union retained large military establishments. Inevitably this led to the emergence of powerful bureaucratic and economic forces which sought to resist any decline in these establishments and, of course, to promote their expansion. It is not possible to quantify these forces, only to describe them briefly in general terms, but it is reasonable to suppose that they have indeed exerted an influence in maintaining and increasing the level of armaments.

An example of such economic pressures is the fact that many politicians in the United States can interpret a decision on whether or not to vote for the procurement of a particular weapon system in terms of substantial changes in the level of employment in their electoral districts. Obvious pressures against any reduction in the level of employment are therefore translated into pressures for the continued production of weapons. Similarly, as an example of bureaucratic pressures, it is perfectly natural for the military establishment, operating as a unified bureaucracy, to attempt to preserve its relative status in the government programme as a whole. Even at lower levels, the various components of the military establishment will attempt at least to preserve, and if possible to expand, their relative status.

Another important factor is the view that, because of the sheer complexity and variety of modern specialized weapon systems, only the military establishment itself is competent to decide the size and character of the national security effort: until recently this view apparently prevailed, at least in the United States. The fact that weapons are both complex and exist in bewildering variety can also be exploited to support claims for higher military spending. Thus, in the United States, debates on the defence budget rarely consider broad questions such as the adequacy of force levels in relation to the requirements of national security and international commitments. Rather these debates get lost in a mass of detail on particular new weapon systems or particular improvements to existing weapons and in equally narrow comparisons with Soviet capabilities, when such comparisons can be made, in order to show that the latter is ahead. This piecemeal review of the military programme contributes to the inflexibility—at least as far as reductions are concerned—of the overall size of the programme.

In summary, the action/reaction process, in its basic form, would seem at best only a partial explanation of the continuing rise in military capability. Other, probably more valid, explanations are to be found in the dynamics of the technological process itself, in the domestic factions that would benefit in some way from the military and in the broader political—mainly foreign-policy—objectives that are believed to be advanced by the maintenance of military forces. No attempt has been made to explore the ramifications of the role which armed force is given in the conduct of foreign relations. It may be observed, however, that in many cases, the political utility of military power is not measured against the requirement of 'defence' proper or the achievement of well-defined military objectives. Instead, military power, measured in crude terms, is considered to contribute to the bargaining power of a country in its international dealings. In this light, it becomes an attractive commodity even in the absence of a plausible military threat or significant risk of armed conflict.

2. *The technological arms race*★

The arms competition which lies behind the 5–6 per cent a year growth in military expenditure does not, for the most part, take the form of a multiplication of existing weapons. There has not been a 5–6 per cent a year increase in world stocks of military planes, submarines or aircraft carriers. It has rather taken the form of a very rapid rate of what is called in civil life 'product improvement': a constant improvement of existing weapons; a very rapid rate of innovation; and a constant search for new potential environments in which weapons can be used. The arms race is now largely a technological one.

There are from time to time sharp quantitative outbursts of increases in the numbers of new weapons. The world stock of intercontinental ballistic missiles, for example, has risen from 500 to 2 650 in the seven years from 1962 to 1969; and there has been a huge increase in the stockpile of nuclear weapons. This can occasionally happen for more conventional weapons as well; the helicopter is hardly a new weapon, but the discovery of new tactical uses for it has led to a three-fold increase in the stock held by the US Army in the last eight years.

But for other important groups of weapons, the actual numbers in the world have tended to stay the same or indeed sometimes to fall: the increase has been in their capabilities.

The example of improvement in tactical air capability

A good example of the way in which arms competition has taken the form of product improvement rather than a simple increase in numbers is provided by the development of the United States tactical air force over the last seven years. Over that time, the number of aircraft in the inventory has stayed about the same: but there has been a very considerable increase in capability. The testimony of the Assistant Secretary of Defense Alain Enthoven in June 1968 makes the point, even allowing for the fact that some of the figures which state the order of magnitude of the improvement are deleted.

'Since 1961, there has been a very substantial increase in our tactical air capability despite the fact that we have about the same number of aircraft today as we had then.

'First, we have more than doubled the payload-carrying capability of the force . . .

'Second, we have developed greatly improved nonnuclear bombs and missiles . . .

'Third, we have developed and are developing automated systems that should improve weapon delivery accuracy. Again, even modest improvements in accuracy result in large improvements in target destruction

★ From *SIPRI Yearbook*, 1968/69.

capability. For instance, reducing the mean bombing error from [deleted] feet more than [deleted] the probability of kill against a tank.

'Fourth, we have developed an ability to seek out and attack tactical targets with conventional weapons at night without flares and in bad weather —a capability we did not have in 1961. [Deleted]

'Fifth, we have vastly improved our ability to deal with the enemy surface-to-air missile and radar-directed anti-aircraft gun defense network . . .

'Finally, we have also increased our investment in sophisticated air-to-air missiles for our fighters. In 1961, only 15 per cent of our fighters carried radar missiles intended for all-weather use. Today [deleted] per cent of them do.'

Research and development comparisons

Behind this extremely rapid rate of technological improvement in weaponry, so much faster than that of civil goods, there is an enormous disparity between the two fields in research and development. For every $100 of military procurement in the United States, Britain and France, there is over $50 of research expenditure. For the general run of manufacturing, the research input for every $100 of output ranges from $1.9 (France) to $7.5 (United States). The disparity is not as great in other countries, but it exists everywhere.

Further, the military research figures are understated. They exclude space research and atomic energy research, both of which have extensive direct military applications. Making some allowance for this, there is little doubt that the research input per unit of output is at least twelve times greater in the military field than in the civilian field, taking the United States, Britain and France together. It is not surprising, therefore, that the rate of innovation and of product improvement is so much higher in military than in civil goods.

This tremendous research and development drive behind the advance in weaponry has an impetus of its own. Once massive funds are voted for weapons research, and once there are large permanent establishments doing nothing but weapons research, it is inevitable that further improvements will be made and inevitable that new fields of warfare will be explored.

*3. Military research and development**

Even though the nuclear weapons stocks of the United States and the Soviet Union already incorporate a large measure of 'overkill', the technological arms race goes on. There is no evidence of any reduction of the enormous military research and development efforts in these two countries. Nor are these vast efforts confined to the

* From *Resources Devoted to Military Research and Development*, SIPRI, 1972, and *SIPRI Yearbook*, 1973.

improvement of nuclear weapon systems: technological advances in 'conventional' weapons as well continue at an unabated rate.

The post-World War II rise in world military expenditure has been accounted for, to a large extent, not by a traditional proliferation of quantities of armaments, but by the technological arms race. This race involves the continual replacement of 'obsolete' weapons by generations of successively more and more complex, costly and lethal types. Research and development (R&D) efforts form the core of the technological competition: they are the source of the continual 'improvements' in weaponry. If all R&D efforts were abruptly stopped, the manufacture of weapons for replacement purposes might still continue; the quantities of existing weapons might be increased; and the arsenals of the smaller and middle powers might even be brought technologically closer to those of the USA and the USSR through international arms transfers. But in its most important form—involving the continual production of new, more dangerous and ever more expensive weapons—the arms race would come to a halt.

Data and estimates

In the course of the study, an effort was made to obtain series of world-wide military research and development expenditures. The data obtained which permitted the construction of estimates of annual military R&D outlays over a period were limited almost exclusively, however, to Western industrialized countries. No information on military R&D funds is published in the Soviet Union or other East European countries, and it has not been possible to provide series of annual military R&D expenditures for any of these countries. Among non-industrialized countries, yearly estimates have been obtained only for India (Table 6.1).

The summary of main results which follows provides a synthesis of the two types of estimates: those based on R&D expenditure data and those derived from examination of the level of total military spending and of the amount of weapons development work under way during the 1960's.

A rough estimate of the annual level of world military R&D expenditure during the 1960's is suggested by this study; the figure comes to about $15–16.5 billion. A handful of industrialized countries with high military expenditures dominate the world's military research and development activity. It is estimated that during the 1960's, the United States and the Soviet Union alone accounted for about 85 per cent of world-wide military research and development. It is uncertain exactly how large the military R&D effort of the Soviet Union is, but it seems to be of the same order of magnitude as that of the United States. There are probably several other countries among those for which expenditure series are not available, which have been spending increasing amounts on military R&D in recent years. Israel and Yugoslavia appear to have the most ambitious weapons development efforts of the countries in this group.

Table 6.1. The level of current military research and development expenditure in 22 countries

US $ mn, current prices, official exchange rates

Countries ranked by level of average annual expenditure, 1967–70 A	*Average annual expenditure 1967–70*[i] B	*Current expenditure 1970*[i] C	*Average annual expenditure 1967–70, at approx. R&D exchange rates*[j] D
United States	8 708.9	8 608.6	8 708.9
France	601.6	536.7	770.8
United Kingdom	583.3	544.8	859.6
FR Germany	271.0	314.2	352.3
Sweden	92.6	74.4	104.4
Canada	79.6	80.8	89.0
Australia	51.0	54.9	–
Italy	22.4	30.5	33.1
Japan	20.8[a]	25.3[b]	52.8[a]
India	18.8	24.4	62.0
Netherlands	12.1	13.9	17.1
Switzerland	7.0[c]	7.7[d]	(9.8)[c]
Norway	5.0[a]	5.5[b]	7.3[a]
Belgium	2.4	2.8	3.9
Spain	–	1.2[f]	(2.4)[f]
Denmark	0.5[a]	0.5[b]	0.7[a]
Greece	–	0.5[g]	1.2[g]
Austria	0.3	0.4	0.5
Turkey	–	0.1[f]	0.3[f]
Finland	0.1	0.1[e]	(0.1)
Ireland	–	0.0[e]	–[e]
Iceland	0.0[h]	0.0[h]	0.0[h]

[a] 1966–69. [b] 1969. [c] 1966–68. [d] 1968. [e] 1967. [f] 1964. [g] 1962. [h] Iceland has no national defence establishment, and no defence or military expenditure of any kind. [i] Throughout this section, the years refer to fiscal years which coincide with *or begin during* the year indicated in tables and text. For countries in which the fiscal year begins on 1 April or 1 July, '1970' refers thus to the fiscal year running from 1 April 1970 to 31 March 1971, or 1 July 1970 to 30 June 1971. The years 1967–70 refer to the four fiscal years 1967–70 (or 1967/68–1970/71) inclusive. [j] The use of special exchange rates is intended to allow for international differences in the costs of R&D input. The rates used here, which are based on manpower and expenditure data for R & D performed by business enterprises, make some allowance for international differences in the wages of R&D employees. They are not specific to military R&D and do not allow for international differences in the prices of R&D facilities and equipment, or in productivity. For further details, see Appendix A.

With the exception of the Soviet Union's effort, there is comparatively very little weapons development under way in the countries for which military R&D estimates are not available. Only four countries in this group had an average annual level of total military expenditure above $700 million during the 1960's: the People's Republic of China, Poland, Czechoslovakia and the German Democratic Republic (GDR). For the GDR, no evidence of weapons R&D was found; and for the other three countries, the range and quantity of projects under way in 1960–68 is small in comparison with that in other countries with comparable levels of military expenditure. All of the remaining countries in the world for which military R&D estimates are not available have had an average level of total military outlays below $500 million—and, by implication, a potential for very limited military R&D undertakings only.

A new phenomenon

Very large government-financed military R&D efforts are essentially a product of World War II, and a new phenomenon in the post-war period. Of course, to take a very long perspective, inventions in armaments have been made since ancient times; and at various periods and in different countries, scientists and scientific advisers concerned with warfare and weapons have been employed by heads of governments and, more recently, by business firms. By 1900–39 a limited amount of more systematic government-financed 'R&D' work was under way on, for example, military aircraft, submarines, tanks and/or chemical weapons in Japan, the United States, some South American countries, and the European powers. Judging, however, by available figures for the United States for the 1920–35 period, when military R&D funds amounted to about $4 million annually, these efforts were very small indeed, set against post-war programmes. In the United States, military R&D spending in 1955 (around $3 billion) was almost 1 thousand times larger than in 1935.

Over half of all government funds allocated to R&D in the United States in 1969 went to military R&D. The comparable proportions in the United Kingdom and France were about 40 per cent and 35 per cent, respectively.

Military research and development—the improvement of existing weaponry and the design and development of new weapons—currently absorbs about $20 billion annually and occupies the time of about 400 000 scientists and engineers throughout the world.

The United States and the Soviet Union

Two aspects of the military R&D activities of the United States and the Soviet Union, related to the very large size of their programmes, are unique and are of overriding importance in the context of international comparisons. The first is their work on the development of strategic nuclear weapons. While the United Kingdom, France and probably China all have operational nuclear weapons, only the USA and USSR have nuclear forces capable of wiping out a large portion of the world's population. For this reason, and to the extent that the state of the strategic balance has world-wide political repercussions, the continuing technological development of US and Soviet strategic forces is of paramount importance.

The second significant aspect of US and Soviet military R&D activities is the extent to which they constitute the spearhead of world-wide technological advances in all types of armaments and military equipment, conventional as well as nuclear. Increasingly since the end of World War II, a 'technological imperative' has come to prevail in the military field, whereby continuing major improvements in the performance of weapons and equipment are taken for granted by political leaders and military planners. The technological arms race, in which increasingly expensive and effective weaponry is regularly brought into the arsenals to replace 'obsolete' stocks, now involves almost all countries in the world—though it is true that most participate as importers rather than designers and producers of the new equipment. Judging by the rate of investment in R&D and by rising procurement and operations and maintenance costs, the rate of technical innovation in armaments is considerably faster than that for almost any civil product. As the countries with by far the largest conventional weapon R&D programmes, and the largest efforts in the area of basic research in military-related science and technology, the United States and the Soviet Union have probably done most to produce the very high rate of innovation in conventional arms. They also contribute most to exploiting new areas of science and technology, new environments and techniques, for military purposes.

Other countries

There are a number of ways in which the military efforts of countries other than the United States and the Soviet Union are significant. First, they contribute, in varying degrees, to the continual world-wide advancement in weapon technology which constitutes the technological arms race. The activities of the industrialized countries in the conventional weapon area, involving work at the forefront of technology, contribute directly to the technological race in conventional arms, perhaps roughly in proportion to the overall size of the efforts. In contrast, the nuclear weapon R&D activities of France, the United Kingdom and China and the conventional weapon R&D activities of the less industrialized and non-industrialized countries, which lag behind the latest developments in technology elsewhere, support the technological arms race only indirectly. These activities do not help to advance the frontier of weapon technology, but, like the import of increasingly advanced weapons by many countries, they do contribute to the general rise in the level of technology incorporated in the weapon inventories throughout the world.

Second, the R&D programmes of a number of countries, particularly those involving activities which lag behind the forefront of weapon technology, may have political implications of special interest. The efforts of some of the less industrialized and non-industrialized countries, such as China, India, South Africa and Brazil, to the extent that they are successful and lead toward the establishment of a self-sufficient defence industry, may contribute to altering relations between these countries and the main industrialized weapon suppliers. In doing so, they may also affect the broader pattern of relations between the industrialized countries and the third world. Certain R&D activities may change regional power balances. The French and Chinese nuclear weapon programmes are of particular significance for the European balance of power and the East-West confrontation in the first case, and for the balance in the Far East and the Sino-Soviet and Sino-American confrontations in the second case. The continued development of sophisticated conventional weapons by Israel and India, to take another example, may have important implications for the conflicts in the Middle East and the Indian sub-continent, and for the involvement of the big powers in these conflicts. New weapon development efforts on the part of Warsaw Pact countries may reflect evolving political relationships between these countries and the Soviet Union.

Third, the military R&D efforts of Western industrialized countries are important because they determine the countries outside the USA and USSR which will be the world's main weapon producers during the next 10–20 years. Within the next decade, it is likely that the United Kingdom and France will be joined by other countries—specifically Japan and West Germany—in this role. In addition, several of the countries with smaller military R&D efforts may expand their defence industries and undertake more cooperative work with the main weapon producers, narrowing the present gap in indigenous weapon development capability between the main producers and the remaining countries.

4. *Escalating armaments**

Military expenditures are useful indicators for an examination of the dynamics of armaments. The long-term trend in military expenditure will reflect the trend in the lethal power of weapons. In the short-term, however, the correlation need not be strong. Nuclear weapons, for example, produced an enormous rise in lethal power but with no corresponding increase in military expenditures.

During this century the quantity of resources devoted each year to armaments has increased twenty-five-fold. Military expenditure has not increased steadily. World Wars I and II were preceded by intense arms races and, therefore, by rapid increases in world military expenditure. But, in each case, the prewar maximum level of expenditure became the postwar minimum level. Thus, world military expenditure since World War II has never fallen below the 1938 level. By that year, the arms race had tripled the 1933 expenditure.

Since World War II expenditure has generally grown rapidly during a war or a major crisis. On each occasion, this has been followed by a period of relatively stable but substantially higher levels of expenditure. There was a large rise after 1948 due to the rearmament programme following the establishment of NATO and the costs of the Korean War. Expenditure then

* Excerpts from Frank Barnaby: 'The Dynamics of Armaments,' *International Social Science Journal*, Vol. XXVIII, No. 2, 1976.

fell somewhat and continued on a plateau until 1960. Another sharp increase came when the so-called 'missile gap' led to the large-scale deployment of land- and sea-based ballistic missiles by the United States. 1964 and 1965 were years of stability but these were followed by a 30 per cent rise due largely to United States intervention in Viet-Nam. In 1970 the level of military expenditure staiblized again and has remained roughly constant.

An analysis of the military spending of the twelve countries which together account for 75 per cent of world military expenditure shows that about 30 per cent is used for uniformed military personnel. Another 30 per cent goes to procure weapons and equipment and a further 30 per cent to operate, maintain and construct the weapons. The remaining 10 per cent is used to fund military research and development (R&D).

Perhaps surprisingly, nuclear weapons account for a comparatively small fraction of the total resources devoted to military uses. The cost of development, production and operation of nuclear weapons and their delivery systems accounts for no more than about 12 per cent of the total world military expenditure.

Military research and development

Historically, the most striking feature of the breakdown in military expenditure is the share absorbed by research and development. Between the two World Wars, R&D absorbed less than 1 per cent of the military budgets of the major powers. But after World War II R&D expenditures sharply increased so that the shares of the military budgets of the major powers devoted to R&D reached 10 to 15 per cent by the late 1950's. They have remained at this level ever since.

A substantial military R&D effort has inevitable ramifications for other categories of military expenditure. The R&D results in a continuous flow of new and improved weapons. Most of these weapons will be produced and deployed. And economics dictates that they be produced in relatively large numbers.

The prime military consideration is not the cost of weapons but their performance. The performance of a given weapon must match or preferably exceed that of the corresponding weapon possessed by the potential enemy. To secure superiority—and to minimize the likelihood of early obsolescence—a new weapon system is often required to have performance characteristics that are unattainable with existing technology. New technology has then to be developed within a prescribed time—inevitably a very costly procedure.

Almost incredible advances in weapons—both nuclear and conventional—have been made over the past three decades. In fact, it is hard to imagine any circumstances in which faster progress could have been made—the only restraint on military technology has probably been the amount of human ingenuity and skill available in the American and Soviet societies. Military technology certainly has not been hindered by arms control and disarmament negotiations or treaties. Nor is it likely that higher funding would have resulted in significantly faster progress. The military technologists have been (and are being) given as much money as they could effectively use.

Advances in military technology depend on military research and development. If this activity were stopped no significant new weapon systems would emerge. But once new weapons are developed, very strong political pressures build up, within the USA and the USSR, for their deployment in large numbers. These domestic pressures are generated by groups interested in maintaining and expanding military and defence industries and establishments. Until these groups are brought under control by strong political leaderships the arms race is unlikely to be controlled, let alone reversed. Bilateral and multilateral arms control negotiations are unlikely to succeed until this internal domestic problem is solved.

5. *The dynamics of military technology**

At no time in man's history has his political organization and the affairs between States been so heavily influenced, if not guided, by military technology, by the weapons of war, as in the years since the end of the Second World War.

The nuclear arms race, the warheads and the delivery systems for them are clearly the most significant elements concerning the major powers in the post-war years. However, there have also been extensive developments in non-nuclear weapon systems since the end of the Second World War. Some examples are:

1. Extensive computer-controlled air-defence networks with large, early-warning, over-the-horizon radars for ballistic missile warning and forward emplaced radar networks for anti-aircraft defence. Smaller fields, mobile, radar-controlled and anti-aircraft missiles, with ranges as high as seventy miles, and analogous shipborne weapons. Self-guided target-seeking air-to-surface missiles.
2. Avionics: Electronics and air-borne computers play a nearly complete role in advanced combat aircraft: navigation, reconnaissance, bad-weather operations, engaging opposing aircraft, fire control, weapons guidance. Airborne anti-submarine warfare has undergone enormous development, with long-range, long-duration patrols, expendable sonobuoy systems, other buoy telemetry, airborne dipped sonars, infra-red and magnetic anomaly surveillance.
3. Increased deployment capabilities: first the airlift, now rapidly deployable air bases. Greatly increased use of the helicopter. Computer-aided logistics management.

* Shortened version of a paper by Milton Leitenberg from *International Social Science Journal*, Vol. XXV, No. 3, 1973.

4. Advanced weapon guidance, using lasers for targeting of many kinds of ordnance in field weapons and ground-support aircraft. Night-time target acquisition and fire-control devices. Radars for artillery and mortar location. The development of 'cluster' dispersable anti-personnel weapons of numerous sorts.
5. Nuclear power for naval vessel propulsion.
6. Chemical warfare: further developments were made by the United States and USSR of the nerve gas agents which had been developed in Germany in the Second World War.
7. The military use of space: satellites are used for military communications, for photographic reconnaissance in visible, infra-red and ultra-violet spectra, for electronic intelligence and for navigation and guidance.
8. The oceans have also seen and are yet to see further invasion by entirely new military-support systems.

An ungovernable process

Sir Solly Zuckerman, for many years Chief Science Adviser to the British Government, wrote:

> The decisions which we make today in the fields of science and technology determine the tactics, then the strategy, and finally the politics of tomorrow.

Not only tomorrow, but of decades ahead. This is seldom realized as clearly as it should be.

National preoccupation with extensive weapons programmes and military might tend to pre-empt political decisions. Military alternatives are available and are recommended for complete satisfaction of 'the national interest', instead of what would perforce be necessary in their absence, a diplomacy of accommodation or compromise.

The vast research establishments develop some new weapon or technique of warfare over the substantial periods of five to ten years. It is automatically assumed that 'the enemy' is likely to discover the new technique as well, and that the only safe thing is to push ahead. In the absence of political constraint, weapons research and development becomes an ungovernable and self-propagating process. It does not need an increase in international tension and hardly a real enemy to keep going.

It is important to understand that a decision such as the one to deploy MIRVs in the United States was made before there was any public mention whatsoever of the weapon system, and even before the members of Congress were informed of the characteristics and implications of the system. The public debate followed the decision. In the USSR and often in France and the United Kingdom as well, such decisions are made without public debate at all.

Strategies, the 'military requirements', at least in the Second World War and after, have always followed weapon capabilities, and it could hardly be otherwise. One can hardly have a military strategy particularly dependent on weapon-system capabilities which do not exist.

Requirements for particular weapons are often conceived of in terms of replacement for a weapon currently in use. Emphasis is placed on production time and on technology in competition on the replacement. Replacement systems are nearly always far more complex than their predecessor. Therefore greater capacity both in development and design teams and in production capacity—plant, machinery, and labour—is required to produce the follow-up in the same time span as its predecessor.

Military research and development

Another method of keeping capacity employed is through R&D expenditure devoted to advance the 'state of the art' in a particular technology which may not at the moment be tied to a particular weapons project. By 1965 only 14 per cent of the $16 000 million United States federal research and development programme was conducted in civil-service laboratories, and less than 20 per cent in federal contract research centres and civil-service laboratories together. Of the remainder more than 70 per cent was performed by industry and about 12 per cent by universities. Industrial 'state of the art' is also pushed by the defence industries in order to improve their competitive advantages for defence contracts, all of which hastens further weapon development by producing advances in materials and production methodology. Competitive incentives have at times also led the industries to development work on uncontracted weapons systems and to attempts to sell the weapons and military functions for them to the services. Cancelled projects perform the same function. Between 1954 and 1965 the United States spent some $6 000 million on major projects which never reached the production stage. If one includes missile systems that were in deployment only a short time this increases to $12 000 million. The United Kingdom spent $1 200 million from 1952 to 1970 on projects which did not reach the production stage.

The aggregate costs of some complex weapon systems which have gone through several 'generations' of components is staggering. The United States has spent some $50 000 million on air defence since the end of the Second World War: interceptor aircraft, distant-early-warning radars, airborne radar picket aircraft, surface-to-air missiles, computers, etc. It is estimated that the Soviet Union has spent several times this amount for the same purpose in the same period. The growth of weapons expenditure is not caused by purchasing increased numbers of weapons, but by increased costs of individual weapon systems. A comparison between the research input per unit of output for civil and for military goods indicates that for every $100 of military procurement, United States or United Kingdom research inputs are of the order of $50 to $60, whereas for every $100 of manufacturing output in general, the research input is of the order of $5 to $8.

Contribution of basic scientific research

One major topic remains to be discussed, the contribution of scientific research to military technology and to advanced weapons development.

The United States Army lists some eighty-nine scientific disciplines grouped into twelve categories in which it supports research. The United States Air Force Office of Scientific Research lists some sixty areas of scientific research in which it has performed a 'colonizing' role: these are areas in which it has an interest and in which it subsidizes research, both in the United States and abroad.

There is no longer any distinction whatever between basic scientific research which may have military relevance and that which does not. This is not because science has changed but because the military 'requirements' have. Weapons are now universally dispersed in all environments—space, sea depths, jungle—and new weapons, communications systems, sensors and support equipment involve so many new energy forms and materials that there probably is no area of scientific research that is not now of interest to the military. The answers to questions of how materials and energy will behave in these newer environments into which weapons systems have moved can only be answered by what is clearly basic research. None of the support of basic science by defence ministries is accidental. It is quite rational and purposeful, and its aim is not primarily the support of scientific research *per se*. It must however feed the goose to obtain the golden egg.

Another significant source of support for military research and development eventually derives from programmes in graduate education. In the mid-1960's AFOSR provided at least partially for the research of more than a thousand doctoral candidates in the United States and of many more candidates at the master's level. These students were 'developing their expertise in areas particularly relevant to Department of Defense interest'. In the early 1960's 65 per cent of all research support in academic engineering in the United States came from DOD and NASA, and only about 6 per cent from industrial sources. Over 30 per cent of academic research in the physical sciences is still dependent on the Defense Department, and the patterns of Defense Department support still set the precedents for much of the support from other government agencies. The guiding of programmes in graduate education to supply manpower and to develop expertise tailored to specific military research and development requirements can be assumed to exist in all countries doing defence research.

A somewhat unfortunate aspect is that research with essentially military applications is sometimes carried out in conjunction with international programmes. This has been true for the International Geophysical Year (IGY), or the Caribbean Bomex programme, which is part of the Global Atmospheric Research Program (GARP), as well as for several other programmes.

A dynamic of its own

We have looked at some of the mechanisms of weapons development since the Second World War. We have seen something of its financial costs, and in particular begun to examine the absolute dependence of

N

advanced weapons development on scientific research, and the way in which research programmes are managed and guided in those nations with heavy investment in weapons development so as to provide as much impetus to these programmes as possible. This is a process with a qualitative change since the Second World War. Military leaderships have learned to direct areas of research towards the information needed to provide weapon systems of ever-increasing capabilities.

Military technology is an area in which notions of inevitable continuity are particularly prevalent, that it has 'a life of its own, and its progress in any direction, however antisocial, cannot be impeded'. If so, then military technology is an explicit example of an area in which man has lost control of his social structures.

6. *The origins of MIRV**

One of the most important military inventions of recent years is the multiple independently targetable re-entry vehicle (MIRV) warhead. A single ballistic missile fitted with a MIRV warhead can deliver, very accurately, one or more nuclear weapons to each of several different targets. Systems consisting of from three to as many as 14 individual nuclear weapons on one missile have been discussed in open hearings on the subject. The first deployment of MIRVs on US intercontinental ballistic missiles (ICBMs), and the mere anticipation of MIRV development by the Soviet Union, have already profoundly affected the dynamics of the arms race.

The multiple targeting of MIRV is accomplished by a device formally known as a Post Boost Control System (PBCS), but more commonly called a 'bus'. Initially, the main rocket booster puts the bus on a course that would cause it to impact somewhere near target number one. The bus contains several re-entry vehicles, a guidance and control system and some small rockets to modify its velocity so that it is aimed as precisely as possible along the orbit to target number one. When this has been accomplished, the bus gently ejects one of the re-entry vehicles. Then, while this first re-entry vehicle continues on its course to the first target, the bus guidance system instructs the propulsion units to modify its course so as to put it on an orbit leading to target number two. Another re-entry vehicle is then ejected and the process is repeated until each re-entry vehicle is *en route* to its prescribed target.

Development and deployment

A number of independent and quite disparate military requirements led to several different lines of technological developments, each of which made

* From *The Origins of MIRV*, SIPRI, 1973.

its contribution to the MIRV programme. As time went on, ideas and personnel were interchanged among the various programmes, which resulted in a complex web of technological developments and inventions. Similarly, there were a number of quite different arguments favouring the deployment of MIRV once it had been developed, and apparently several of these would have been sufficient in themselves.

It can be said that the MIRV programme had many roots and branches. Important decisions were made by many people, only loosely connected with each other, and over a period of more than a decade. Of all the stimuli that gave rise to the *development* of MIRV, the most important was the perceived need to penetrate, with assurance, ABM systems whose theoretical capability was slowly but steadily improving with time. However, all the technologies needed for MIRV had other reasons underlying their development, and so MIRV would almost certainly have emerged at about the same time even if the need for ABM penetration had not been perceived until much later, and probably even if it had never arisen at all.

Thus, during the development phase, the MIRV programme was almost entirely technologically determined in the sense that the key decisions were made by technologists who were either attempting to solve problems posed by nature, or responding to their perceptions of the technological challenges posed by the Soviet missile space programmes.

Technological and political factors

Unlike the arguments used during the development phase, those used in the discussion over deployment included both technological and political factors. The decision to *deploy* MIRV was compelled by three factors, all of them independently leading toward the same conclusion. They were:

(1) MIRV was a technological certainty. The invention and testing of all its major components were already accomplished facts, and the internal momentum of the programme was urging it on toward deployment.

(2) MIRV continued to be the most promising way of countering any conceivable Soviet ABM development or deployment. Its deployment thus solved the strategic problem posed by Soviet ABM deployment. By the same token, it became an argument against deployment of ABM by either side by emphasizing the futility of such a move.

(3) MIRV provided a relatively inexpensive way of expanding the number of points that could be targeted. Its deployment thus put an end to the argument over force expansion.

Counterposed to these three factors were the predictions by the ACDA experts about how MIRV would have a destabilizing influence on the arms race. Their arguments turned out to be correct, but they were not the whole story, even from the point of view of arms-control advocacy. Further, the number of people holding such views was small compared with the number of those influenced by one or more of the factors favouring MIRV deployment, and so plans to deploy proceeded inexorably throughout the debate.

The moral in this history for arms-control and disarmament advocates is this: programmes like MIRV, that is, programmes evolving from many independent and seemingly unrelated purposes and decisions, cannot be controlled or stopped by directly confronting them. They can only be slowed or stopped by slowing or stopping the arms race as a whole.

7. The military significance of nuclear tests*

What are the military attractions to the USA and the USSR of continuing nuclear tests?

Nuclear tests can be classified in four categories as follows:

1. *Confidence tests:* occasional tests for the purpose of maintaining confidence in weapons already stockpiled. Since materials age, inadvertent changes may occur in production, and other things can go wrong, military authorities frequently insist on sampling the performance of their weapons. These tests are not intended to advance the state of the art.

It is not clear that such tests are really needed since a bomb has no movable parts and it is possible to test the operation of the whole device short of 'going nuclear'. Nuclear fuels work, as the world knows too well, and their chemical integrity can be checked by conventional chemical means. If all tests were prohibited, steps would surely be taken to minimize the possible deterioration in weapons reliability. Bomb designs, material standards, production methods and so on would be rigidly frozen.

2. *Proof tests:* tests of newly designed weapons where the design is based on accepted and established principles. New designs are needed to meet size, weight, shape or other performance requirements. It is necessary to test before stockpiling a new weapon to see whether indeed the new model works as intended and expected.

Without proof tests it would be necessary to design new weapon systems around existing nuclear-bomb designs. The effect is different for weapons for deterrence and weapons to enhance war-fighting capability.

For the purpose of maintaining a credible deterrent, the case for continued nuclear testing is hardly persuasive for several reasons: (*a*) Deterrent weapons tend to be in the larger yield class where warheads of high efficiency are already available to both the USA and the USSR. (*b*) Several different bomb designs can be used to make up a deterrent force, thus minimizing the possibility of a catastrophic failure in its retaliatory capability. (*c*) So much overkill already exists that greater efficiency in the nuclear weapon arsenals of either power can hardly add to its deterrent capability.

For weapon systems intended for war-fighting, the value of continued testing is more debatable. In this case, weapon systems are designed to deal

* From *SIPRI Yearbook*, 1972.

with other weapon systems, be they surface-to-air missiles, tanks, or field armies, and superior performance may make military sense. Technical improvements in nuclear-warhead design can make for superior weapon systems. On the other hand, it can be argued that the USA and the USSR, with their long history of intensive nuclear research and development and with their many weapon tests, undoubtedly have a very complete warehouse of weapon designs already, so that a suitable, though probably not optimal, weapon design can be found to fit almost any particular requirement. Moreover, a nuclear weapon system for war-fighting purposes consists of much more than a bomb. Other technical aspects of the system (for example, aircraft speed and agility, tank speed and range), in addition to such factors as military tactics and troop training, will generally be more important to the performance of the system than the precise character of the nuclear bomb it delivers.

3. *Effects tests:* nuclear explosions are used to provide a realistic nuclear condition in which to test materials, electronic devices, the survival of weapons against defence measures, and so on.

Tests of this kind are now conducted underground. The USA and the USSR by now have had so much experience with underground tests that they could pursue a serious programme of effects tests restricted to yields below the seismic detection threshold.

4. *R&D tests:* this includes tests to investigate entirely new principles in weapon design. Such tests would be needed to advance the state of the art towards laser-initiated pure fusion bombs, neutron bombs, or major advances in yield-to-weight ratios for very small weapons. Large weapons are already very close to their theoretical maximum in yield to weight, and improvement by a factor of about two is all that we can reasonably expect. That is quite insignificant compared with the advances made to date: the yield of current weapons per unit of weight is about 1 000 times greater than that of the first Hiroshima bomb.

Vigorous research and development appears to be continuing on the assumption that a breakthrough, comparable in significance to the original development of nuclear weapons and thermonuclear weapons, may lie around the next scientific corner.

Research and development can continue without nuclear testing, but, of course, with a greatly reduced scope.

8. *Strategic doctrines of NATO and the WTO**

There are three clearly discernible periods in the evolution of the strategic doctrines of the two alliances. In the first few years immediately

* From *SIPRI Yearbook*, 1974.

after World War II, the doctrines of the United States and the Soviet Union were influenced primarily by the nature of the military forces available to each. The United States had, in addition to strong ground and naval forces, a monopoly of nuclear weapons, while Soviet strength was confined to large numbers of ground troops. As a result, the United States, on the one hand, relied on nuclear weapons to deter a westward expansion of the overwhelmingly superior conventional Soviet forces into Western Europe, and it extended the 'nuclear umbrella' to this region through the creation of NATO in 1949. At the same time, it contributed to the re-equipment of West European forces and the re-establishment of West European defence industries. The Soviet Union, on the other hand, relied on its conventional superiority in the European theatre to deter Western aggression, including the possibility of surprise nuclear attack; following an immediate postwar demobilization of the majority of its troops, it doubled the number of forces in the period from 1948 to 1955. It also pursued the development of nuclear weapons intensively, exploding its first atomic weapon in 1949, and its first thermonuclear device in 1952. In addition, following the entry of the Federal Republic of Germany into NATO and the initiation of West German rearmament in 1953 and 1954, the Soviet Union undertook the establishment of the WTO, in an attempt to consolidate the eastern front. In sum, this first period may be considered a time of transition, marked by a considerable fear to attack on both sides; a movement toward the establishment of the two alliances; and a reliance on existing advantages in military strength.

The second period—from the early 1950's to the early 1960's—is characterized by increased reliance on nuclear weapons on both sides, including, for a brief period on each side, exclusive reliance on nuclear armaments. This began with the adoption of the doctrine of massive retaliation by the Eisenhower administration in the United States. At the time this doctrine was most closely adhered to—in the mid-1950's—Soviet conventional forces were at their peak (in quantitative terms). Soviet development of nuclear weapons was proceeding rapidly and the Warsaw Treaty Organization was newly formed. It had also become clear that NATO would not actually attempt to match the WTO—or even the Soviet Union alone—in the number of conventional forces deployed in Europe. However, the United States retained a substantial advantage—an estimated ratio of about 10:1 in the number of nuclear weapons, as well as a monopoly of the means of delivery of these weapons to the territory of the opposing power (in the form of the earliest long range and forward-based medium-range bombers). A doctrine which threatened all-out nuclear retaliation as a response to any sort of aggression on the part of the enemy appeared a suitable US posture—which was both credible and appropriate with regard to the existing balance of forces. The Soviet Union, while moving towards the development and deployment of a credible intercontinental nuclear force, had no alternative but to rely on its conventional superiority in Europe in the early 1950's. However, following the introduction of its first intercontinental bombers in 1956, the USSR was able to place some emphasis on nuclear deterrence and to begin a reduction of some 50 per cent in the size of the armed forces over the period from 1955 to 1960. A debate on the respective roles of nuclear and conventional forces, pursued in the Soviet Union in the late 1950's, was brought to an end in early 1960 when Khrushchev announced the adoption of a new nuclear policy, the successful production of intercontinental ballistic missiles and a planned further reduction in the size of the armed forces. In both the United States and the Soviet Union, the announcements of virtually complete reliance on nuclear weapons for the prevention of a major confrontation were soon followed by some retraction, allowing a continued role for conventional forces. For both powers the second period is, nevertheless, dominated by the idea of nuclear deterrence.

The third period, extending from the early 1960's to the present time, has been marked in both East and West by a rejection of the nuclear-oriented doctrines of the 1950's, and an evolution of more flexible strategies. Again, it was the United States which took the first step in this direction. Factors which probably contributed to the development of the Kennedy–McNamara doctrine of 'flexible response' in the early 1960's, and the eventual adoption of this doctrine by NATO in 1967, include: the lack of credibility in the 'all or nothing' alternatives posed by the doctrine of massive retaliation; reaction against the dangers of a holocaust raised by this doctrine; and the steady build-up of Soviet strategic nuclear forces during the 1960's, in a manner which would eventually lead to nuclear parity between the two sides. The doctrine of flexible response implied greater reliance on increased conventional forces. In addition, it involved greater emphasis on tactical nuclear weapons.

President Nixon's doctrine of realistic deterrence and recent statements by Soviet military theorists have confirmed the common trend away from exclusive reliance on either nuclear or conventional forces and towards the development of a broad spectrum of armaments and support for a more complex strategy calling for the use of forces specifically appropriate to different types of conflicts.

7. Economic and social consequences of armaments

1. *UN report: economic and social consequences of military expenditure**

In October 1971 the United Nations Secretary-General presented a report on the economic and social consequences of military expenditure to the General Assembly, prepared and unanimously adopted by a panel of fourteen experts. The report is summarized below and then commented on.

World military expenditure in 1970 amounted, roughly, to $200 billion. This was, at constant prices, a rise of the order of $50 billion, or a third, since the beginning of the decade. To indicate an order of magnitude, $200 billion is roughly what the governments of the world spend on health and education.

So large an expenditure on military preparations in a period of comparative peace is a new phenomenon. In the period before World War I, for example, 3–3½ per cent of world national output went to military spending. The figure is now 6–6½ per cent of an enormously increased output, so that in real terms military spending has risen at least twenty-fold in the last fifty years.

The developing countries account for 7 per cent of the world total. However, their expenditure has been rising faster in the developed countries in the last decade. This is largely accounted for by the countries in the Middle East and Far East; and the conflicts there have not, of course, been independent of the confrontation between the great powers.

There are particular ways in which the 'share of national product' measure understates the impact of military spending. First, military uses absorb a disproportionately large amount of scientific and technological ability in military research and development. (The report roughly estimates that military research and development probably absorbs some $25 billion of an estimated world total research and development expenditure of some $60 billion.) Secondly, high levels of skills are needed for the operation and maintenance of sophisticated weapons. Thirdly, military expenditure is a charge on government revenue, and therefore sets a particular constraint on government social expenditure. Fourthly, military spending has been a disturbing factor in the world balance of payments.

Although the figures of military expenditure may have been rising rather more slowly in the sixties, a study of the weapons development of

* From *SIPRI Yearbook*, 1972.

that period suggests that the technological arms race was accelerating. This technological arms race, which provides the main dynamic behind the upward movement of military expenditure, is carried forward to some extent under its own impetus. It is open-ended; in the absence of political action there is nothing to stop it going on indefinitely.

The national consequences

The cost of the arms race consists of the sacrifice of the alternative uses to which the resources might have been put.

The resources might have been used to accelerate the rate of economic growth, either by being used to increase fixed investment, or to give more resources to 'invest in man'—the improvement of man's knowledge, skill, and organizational ability. The standard methods of calculation suggest that a transfer of resources to these purposes could have made a perceptible difference to growth rates.

Military expenditure has not only absorbed considerable resources; it has been a destabilizing element in the world economy. It cannot easily be adjusted to the economic situation of the country because it tends to establish a prior claim and has frequently set off inflationary spirals.

The balance-of-payments effects are felt particularly by developing countries, which have to import all the more sophisticated weapons from the industrial countries; this can pre-empt a good part of their earnings of foreign exchange.

The one significant benefit claimed for military expenditure is the civil spin-off from military research and development. However, if the same resources were devoted directly to civil problems, a much larger return in usable civil technology could obviously be obtained. Further, military research and development has tended to distort the whole pattern of world research and development, away from the pressing needs of agriculture, population and medicine.

The social consequences of military expenditure are harder to establish categorically. But there is little doubt that the spectacle of a world increasing its already absurd levels of lethal power while neglecting the urgent welfare needs of billions of people has been one factor in the disillusionment of many young people.

The military-industrial complexes to which heavy military spending gives rise inevitably exert influence on policy, towards maintaining the circumstances and conditions which gave rise to them. (The change in the traditional relationships between the civilian and military sections of the economy can result in a threat to democratic processes.)

The international consequences

The arms race and international tension are inextricably intertwined. In many cases, international tension can certainly lead to arms races; but arms races in turn exacerbate that tension.

Nations which have spent large sums on armaments will be predisposed to look for military solutions to their disagreements.

Because of the very high research and development costs of modern weapons, medium-sized powers will seek to sell arms to third world countries in order to maintain a viable defence industry. There is also a strong economic incentive to sell surplus weapons as weapons rather than as scrap.

There are also a number of links between military expenditure and world trade. Strategic embargoes are a feature of a heavily armed world. The restrictions which still remain apply to a number of commodities which are of key importance in modern industrial and engineering development. It is not only trade that is affected: the exchange of knowledge and technical 'know-how' is also impeded.

The level of trade between the developed, market economies and centrally planned economies is much lower than might be expected in a fully peacetime world; the armed tension between the two groups of countries is no doubt one reason for this.

Perhaps the main effect of military expenditure on aid to developing countries is to put it lower in the list of national priorities. Certainly aid to developing countries is extremely small—only one-thirtieth of the world total of military expenditure.

There is good reason to think that if world military expenditure had been substantially lower, aid would have been higher.

Military considerations have to some extent distorted the direction of aid; economic aid has followed closely the pattern of military aid, which would hardly be the case if it were properly serving the requirements of world economic development.

Military expenditure in developing countries themselves can be a strong constraint on their economic growth because it pre-empts scarce resources at three points. It takes scarce resources of government finance, scarce resources of foreign exchange, and scarce resources of skilled labour.

At the moment, very few research and development resources are devoted to the needs of developing countries. Since military research and development accounts for a substantial proportion of the world total, there is considerable scope here for a productive transfer—though it would take time.

In general, aid is so small, and military expenditure so large, that a very small transfer of resources would radically transform the aid picture. Five per cent of the world military budget would enable countries to meet the aid targets of the Second Development Decade. Ten per cent of world military expenditure could increase the total of fixed investment in developing countries by a third.

Comments on the report

The main point of a document of this kind is that it is an agreed, unanimous document, signed by scientists from Eastern countries, Western countries and the third world.

The fact of agreement is as important as the opinions actually expressed.

The same point can be made about the report which was the predecessor of the present one: the report on *The Economic and Social Consequences of Disarmament* which was presented to the UN General Assembly in 1962. The most important point about that report was the agreement among the experts that disarmament was technically feasible in capitalist countries: the doctrine that military expenditure was economically essential to them was abandoned. The preface to that report by the Secretary-General emphasizes this:

> It is particularly encouraging that the Consultative Group should have reached the unanimous conclusion that 'all the problems and difficulties of transition connected with disarmament could be met by appropriate national and international measures', and that 'there should thus be no doubt that the diversion to peaceful purposes of the resources now in military use could be accomplished to the benefit of all countries and lead to the improvement of world economic and social conditions'.

The present report is much stronger on the economic consequences of the armaments race than on the social and international political consequences. This was to be expected, given that the experts were to some extent national representatives. It is not really possible for a group of this kind to go deeply into questions which are matters of acute political controversy. So the study does not go into any detail about the effects of the existence of a large military establishment, and its industrial concomitants, on the nature of society or on the values expressed in political discussions, in the mass media, and so on. One reason for the paucity of material presented on this subject was no doubt that virtually all the evidence and discussion on this question uses Western countries as examples. There are no published Soviet studies of the social consequences of the existence of a large military sector in the Soviet Union. Consequently, any extensive treatment of this subject would have appeared biased.

In much the same way, the report treats the international political consequences of the arms race and military expenditure delicately. Here again it was not politically practicable for the group to have any extensive discussion of such matters as the consequences for conflict of the considerable supplies of weapons from the great powers to third world countries.

The main message of the report, therefore, is to serve as a reminder of the immense quantity of resources which the world wastes in preparation for mutual slaughter, and as a reminder of all the massive benefits which would accrue if these resources were devoted to positive ends.

2. *UN report: disarmament and development**

At the 25th General Assembly of the United Nations a resolution was adopted requesting the Secretary-General to submit a report on measures to ensure that the link between disarmament and development be fully understood and utilized in as practical and comprehensive a manner as possible. The report, entitled *Disarmament and Development*, has been prepared and unanimously adopted by a group of nine experts. A summary of the main conclusions and recommendations of the report is reproduced here.

(*a*) Disarmament and development can be linked to each other because the enormous amount of resources wasted in the arms race might be utilized to facilitate development and progress. Furthermore, the blatant contrast between this waste of resources and the unfilled needs of development can be used to help rouse public opinion in favour of effective disarmament, and in favour of the achievement of further progress in development particularly of the developing countries.

(*b*) World military expenditures in 1970 were roughly $200 billion, that is, 6.5 per cent of the GNP of the countries of the world. Military expenditures of the countries which provide aid for development are estimated to be approximately 6.7 per cent of the GNP, or 25 times greater than the official development assistance they provide.

(*c*) The two principal aims of policies of world-wide economic and social development are the increase in the levels of living of all peoples and the reduction in income disparities both within and between countries. On the basis of the modest achievements of the First United Nations Development Decade, and even if growth objectives of the Second United Nations Development Decade are attained, the problem of reducing mass poverty and unemployment in the less developed regions of the world still remains. More efforts therefore should be made by the world community. While in the opinion of the group, developing countries bear responsibility for adopting adequate measures to mobilize their own resources more effectively, and for reducing income disparities, the solution of that problem would, in many developing countries, depend on the contribution to their external resources made by expansion of their exports and also, to a significant extent, on stepped-up foreign assistance.

(*d*) Disarmament would contribute to economic and social development through the promotion of peace and the relaxation of international tensions as well as through the release of resources for peaceful uses. The transfer to peaceful uses of resources used in each country for military purposes will bring about greater satisfaction of civilian needs of the country. The resources thus released, sometimes referred to as the 'disarmament dividend', can be redirected to raise standards of living and to promote faster growth.

* From *SIPRI Yearbook*, 1973.

(*e*) There will be considerable variation among developing countries respecting the magnitude of their own 'disarmament dividend'. In case of general and complete disarmament—and also, to a lesser extent when the cuts in military expenditure are significant but less than total—economic assistance granted by developed to developing countries could and should be greatly increased and should be given higher priority in the allocation of released resources. Since military expenditures now absorb a larger proportion of the combined GNP of the developed than of the developing countries, a general (proportional) reduction in military expenditures will increase the non-military part of the GNP of the first group of countries proportionally more than that of the second group. However, a simultaneous increase in the fraction of GNP in the advanced donor countries allocated to international development assistance could not only prevent a widening of the 'gap', but contribute greatly to its closing.

(*f*) The group suggests that consideration should be given to progress in disarmament in the periodic reviews and appraisals of progress towards achieving the goals and objectives of the International Development Strategy for the Second United Nations Development Decade.

(*g*) Most of the resources released by disarmament, total or partial, would be readily transferable to other uses—for example, manpower, food, clothing, transport, fuel and products of the metal and engineering industries. Budgetary action to raise civil demand will be enough to induce redeployment of these resources either to investment or to consumption, public or private. But other resources—for example, nuclear-weapon plants and military-aircraft and missile plants—may not be readily transferable.

(*h*) The group suggests that governments, when placing orders for specialized military production or creating specialized plants likely to give rise to transfer difficulties in the event of disarmament, should make advance plans to deal with the redeployment to peaceful work of the manpower and plant (in so far as the latter is reusable).

(*i*) Apart from catering for these areas of special difficulty, all countries might be urged to consider what would be the most valuable ways of redeploying resources from military to civil use and to consider, in particular: which specialized resources now used by the military might make a particularly valuable contribution to development in any area; and, in the light of such an assessment, which specialized resources would be suitable as aid or technical assistance from developed to developing countries. Planning of this kind would benefit from international cooperation.

3. Consequences of weapon imports to developing countries[*]

Each year, increasing quantities of resources in developing countries are devoted to the procurement of major weapons from abroad. The gross national products of all developing countries have grown at an average rate of 5 per cent a year since 1950. Their military expenditures have grown at a yearly rate of 7 per cent, while their major weapon imports have grown at a rate of 9 per cent.

The total average yearly expenditure on major weapon imports to developing countries in 1965–70 has been around $1.4 billion. This represents approximately 3 per cent of their total imports. In certain regions, the share is higher. Thus, expenditure on major weapon imports represents 8 per cent of total imports to the Middle East, and 5 per cent of total imports to South Asia. Total imports of military equipment are about twice the size of major weapon imports, and thus represent around 6 per cent of the total imports of developing countries.

The trade in weapons also absorbs other kinds of resources. In particular, the infrastructure and skilled manpower required to maintain a major weapon in operational condition is quite considerable. To repair and maintain a tank in an advanced country, 400 man-hours are required per year. For a destroyer, 45 000 man-hours are required. For each aircraft, four men are required full time for operational maintenance and six men for overhaul. These figures do not include the costs and the skilled manpower requirement for establishing repair shops, nor do they include the various support services necessary for an effective field organization. For example, for an effective field organization, 50 men are required to support each aircraft. If all these costs are taken into account, the total cost of maintaining and operating a major weapon will be several times its initial price. So it is in the use of skilled manpower as well as in the absorption of foreign exchange that imports of major weapons bear heavily on the economies of poor countries.

It is argued that the importing of these weapons provides training which would not otherwise be given, and which is of considerable potential use in civil life. This is part of the general argument about the role of the military in third world countries. How far this is in fact true depends, first, on what alternative provision of training there might be if arms supplies (and military expenditure in general) were lower. Secondly, it depends on the extent to which the skills learned in operating and maintaining weapons can be transferred easily to the civil sector. It is probably true that the more sophisticated the weapons are, the more specific the skills. Obviously mechanics trained to service army trucks can easily transfer to civilian vehicles. However, the skills learned in the maintenance and servicing of aircraft—in a country

[*] From *The Arms Trade with the Third World*, SIPRI, 1971.

where there are hardly any civil aircraft—are probably much less easily transferred.

Equally, it is sometimes argued that the acquisition of military weapons leads to the construction of roads, airfields or other items of 'infrastructure' which might not otherwise be provided. However, the resources which are used to provide them for military use could just as well be employed to provide them for civil use. Further, military roads and airfields are frequently in areas—border areas, for example—where the potential civil use is minimal.

However, it is not very helpful to treat the problem of transferring resources from military to civil uses as a purely mechanical one. Most governments have some leeway in reducing expenditure on weapons and increasing expenditure on development; but it is generally not realistic to envisage large transfers of resources without other political and economic changes. Arms procurement policy cannot be treated in isolation.

For example, in a number of countries in the Middle East and South Asia, heavy arms procurement has accompanied modernization and economic reforms because nationalist groups have been in favour of both. But elsewhere the link may be between heavy arms purchases and a general reactionary posture, as weapons are procured to bolster a feudal regime. What would happen in any particular country as a consequence of a large change in arms supply policy would depend on the way in which this change was associated with other policy changes or changes in regime.

There is another link between arms supplies and economic development, which arises from the role of arms supplies as one element in the hegemonic relationship between suppliers and recipients. This relationship, in turn, influences the economic structure of recipient countries and the economic policies of recipient governments.

4. *The consumption of raw materials for military purposes**

From the point of view of the military consumption of raw materials, two features of the post-war-period are relevant. First, and most important, is the size of the military establishments maintained in peacetime. On the average, annual world military expenditure since World War II has been more than five times as large as the average between the two world wars, excluding the effects of inflation. In fact, the volume of resources consumed annually for military purposes has increased about thirty-fold over the course of this century. At the present time the world diverts about 6 per cent of its total output to military uses. For many years over the post-war period this fraction was even higher, around 8–9 per cent. In contrast,

* Shortened version of a paper by Ronald Huisken from *Ambio*, Vol. 4, No. 5–6, 1975.

before World War I and during the inter-war years before the onset of the arms race that preceded World War II, some 3–3.5 per cent of total world output was devoted to military uses. It can reasonably be inferred, therefore, that the consumption of raw materials for military purposes has increased dramatically. Secondly, whereas the quest for technological superiority has reduced the relative importance of raw materials in the production of weapons, it has also created a rapid turnover of weapons. The design, development and production of weapons systems is now a continuous, indeed overlapping, process. An additional point worth noting is that as weapons become more effective, and particularly if an imbalance prevails between offensive and defensive weapons, losses in time of war will be high. This was vividly demonstrated in the Middle East in October 1973. In just 19 days of fighting the Arabs and the Israelis expended thousands of tonnes of munitions and between them lost about 2 000 main battle tanks and more than 500 combat aircraft.

Some specific data

To be precise on the quantity of raw materials consumed for military purposes is very difficult. Statistics on the worldwide military consumption of raw materials are simply not available. As is so often the case in this field, accurate statistics exist only for the United States. Nevertheless, the USA presently accounts for more than 30 per cent of world military expenditures and, given its predominant position as an arms producer, probably accounts for a significantly higher percentage of the total worldwide military consumption of raw materials. Very similar orders of magnitude apply for the Soviet Union. A rule of thumb could be that worldwide military consumption of raw materials is unlikely to be less than double that of the US.

Table 7.1. Military use of selected raw materials, USA, 1970

US military use as percentage of total use	
Bauxite	14.0
Copper	13.7
Lead	11.3
Zinc	11.0
Nickel	9.7
Molybdenum	9.3
Tin	8.8
Chromium	7.6
Iron	7.5
Manganese	7.5
Petroleum	4.8

Source: S. P. Dresch, *Disarmament: Economic Consequences and Development Potential* (Yale University and National Bureau of Economic Research, New Haven, Connecticut, December 1972), p. 32, Table 4.

Table 7.1 gives, for selected raw materials, the percentage of total US consumption directly attributable to the military. In 1970 the US was still heavily involved in Vietnam, so the percentages are somewhat inflated compared to 'normal' military consumption. For example, by 1973, military consumption of petroleum had declined to 3.7 per cent of total US consumption. But even so it is apparent that the military consumption of these materials is by no means negligible. To give some indication of actual quantities, during fiscal year 1971 shipments of aluminium (bauxite) by US industry totalled about 4.9 million tonnes; more than 0.6 million tonnes was consumed by the military. Similarly, US military requirements accounted for about 249 thousand tonnes of copper. By way of comparison, copper production in China in 1970 was estimated at 109 thousand tonnes.

As the performance parameters specified for weapons systems become more demanding, so the use of special-property materials increases. An example is titanium in the case of aircraft. The F-8 and the F-105, both US combat aircraft produced in the 1950s, had 8–10 per cent of their airframe weights composed of titanium. Present generation aircraft, such as the F-15 Eagle and F-14 Tomcat, have between one-quarter and one-third of their airframe weights composed of titanium And the SR-71, a US strategic reconnaissance aircraft capable of cruising at three times the speed of sound, is constructed almost entirely of titanium and its alloys. It is not surprising, therefore, that in 1972 the estimated military demand for titanium in the US was 4 800 tonnes, about 40 per cent of total US demand for this metal.

With thousands of aircraft and ground vehicles and hundreds of ships, the US military establishment is understandably a massive user of petroleum. Estimated consumption for fiscal year 1974 was 232.5 million barrels after economy measures were taken in view of the oil crisis; in fiscal year 1973 consumption was 273 million barrels. This is more than double the pre-Korean War level of consumption but less than 70 per cent of the level prevailing at the height of the Vietnam War when the US military was consuming in excess of 1 000 000 barrels per day. It should also be pointed out that these figures exclude the petroleum products consumed in the production of weapons and military equipment. Using present US consumption as a basis, annual worldwide consumption of petroleum for military purposes can be estimated at about 700–750 million barrels. This should be compared with 360 million barrels for the whole of Africa and 825 million barrels for South Asia and the Far East (excluding China and Japan).

Resources and armaments

At the broadest level it can be argued that the competitive accumulation of armaments and the importance attached to military strength has distorted the allocation of resources both nationally and internationally. Many countries encourage the establishment and maintenance of defence and defence-related industries to an extent that would not be justified if purely economic criteria were applied. Similarly, the pattern of international trade is distorted by prohibitions on the export to adversary nations of materials and products that may contribute to their military potential.

At another level one can point to the blatant contrast between the resources devoted to armaments and the assistance provided by the industrialized nations to the developing countries. *The diversion of a mere 5 per cent of the combined military expenditures of the developed countries would double the existing volume of official development assistance provided annually to the developing countries.* There is probably no more vivid indicator of the distorted priorities which have prevailed over the post-war period. It should be mentioned here that many under-developed countries are also devoting a large and rapidly growing quantity of their scarce resources to armaments. Table 7.2 shows the trend in the proportion of gross domestic product devoted to armaments in selected developing countries. In reading these percentages it should be remembered that the weighted average for the whole world has been declining for the past several years and is now about 6 per cent. Collectively, the developing countries have increased their share of total world military expenditure from 4.6 per cent in 1960 to 10.8 per cent in 1974.

Table 7.2. Military expenditure as a percentage of gross domestic product: selected developing countries, 1960–73

Country	1960	1965	1970	1973
Egypt	5.6	7.7	18.0	–
Iran	4.2	4.7	7.4	8.7*
Iraq	7.1	9.2	11.1	–
Israel	6.6	7.9	23.6	20.6*
Jordan	19.4	12.8	17.8	16.1
Syria	–	7.9	9.6	15.0
India	1.9	3.6	3.0	–
Pakistan	2.8	4.0	3.7	6.4
Brazil	2.0	2.5	2.2	2.2*
Chile	2.6	2.0	2.6	2.7*
Peru	2.4	2.9	2.9	3.3*
Kenya	0.4	1.0	1.1	1.4
Libya	–	1.4	9.8	8.8*
Morocco	2.3	2.4	3.0	4.0
Tanzania	–	0.8	1.9	2.4
Zambia	1.1	1.8	1.4	5.4*

* 1972. – = not available.
Source: SIPRI Yearbook, 1975.

The grossly uneven pattern of consumption of raw materials is at this moment a hotly debated subject as part of the general endeavour to forge a new world economic order. It can only be a matter of time before it is explicitly recognized that the waste of natural resources by any country is not a loss to that country alone, but a loss to the world community. And, of course, resource consumption for military purposes is the largest and clearest form of waste.

8. Arms control and disarmament

1. *Disarmament efforts 1945–1965**

On 26 June 1945 the United Nations Charter was signed in San Francisco. The new world organization proclaimed as one of its main purposes and principles the maintenance of international peace and security. In order to promote this purpose the founding members entrusted specific responsibilities for disarmament and the regulation of armaments to the Security Council and the General Assembly, thus providing the legal basis for all further activities in this field.

The Security Council was made responsible for formulating, with the assistance of the Military Staff Committee (art. 47), plans to be submitted to the members of the United Nations for the establishment of a system for the regulation of armaments (art. 26). The General Assembly was empowered to consider the principles governing disarmament and the regulation of armaments and to make recommendations about them (art. 11).

When the Charter came into force (24 October 1945), disarmament had already become a serious problem. The dropping of the first atomic bombs (August 1945) confronted the United Nations with the urgent task of establishing control over atomic energy and outlawing the use and production of atomic weapons. Thus the very first resolution passed by the General Assembly (24 January 1946) unanimously established the UN Atomic Energy Commission, consisting of the members of the Security Council, plus Canada (when not a member of the Council). The Commission was asked to draw up plans for the control of atomic energy and for the elimination of atomic weapons and of all other major weapons of mass destruction. In the immediate post-war period disarmament negotiations were almost entirely concerned with these questions.

The United States plan for nuclear disarmament (Baruch Plan) was put forward in the Atomic Energy Commission in June 1946. It envisaged the creation of a system for control of atomic energy, with punishment for violation of the rules of operation. This would then be followed by the stopping of the manufacture of bombs and the destruction of all existing bombs. The plan provided that *control must precede prohibition*: and the administration of the control would be free of the veto of permanent members of the Security Council. The suggested degree of inspection and control—probably to be exercised by a body in which Western powers

* From *SIPRI Yearbook*, 1968/69.

O 2

would have the major influence—was unacceptable to the Soviet Union. Furthermore, if at any time the treaty broke down, the United States would have a monopoly of atomic weapons. The Soviet Union's counter-proposal of June 1946 (Gromyko Plan) required signatories to agree not to use atomic weapons, to prohibit the production of them and to destroy all existing stocks. In this plan, *prohibition and destruction would precede control*: consequently, the United States advantage in atomic weapons would be nullified. The Soviet Union modified this position in October 1948, suggesting that the conventions on the prohibition of atomic weapons and on the establishment of international control over atomic energy be brought into operation simultaneously.

It was some time before discussion on conventional armaments began. The Commission for Conventional Armaments, established by the Security Council in 1947 and consisting of its members, did not begin serious work until 1948, after the rejection of the Soviet proposal that conventional and nuclear disarmament be considered together. The Soviet proposal, which then called for reductions in existing forces by a third, was not accepted because in the Western opinion such reduction would preserve the Soviet Union's conventional military superiority. The Western countries were concerned with the relative level of conventional arms, and most immediately with establishing what the existing ratio in fact was.

In this period, then, when the United States had a nuclear monopoly and the Soviet Union was presumed to be superior in conventional weapons, each side was making proposals which preserved its own position while neutralising the other side's superiority. The two Commissions, failing to agree, adjourned indefinitely in 1950.

Regulation of armed forces and armaments

After two years of deadlock, negotiations began again in 1952. By this time the Soviet Union had exploded an atomic device (1949); and the Western powers agreed to merge the discussions of nuclear and conventional disarmament. A new Disarmament Commission was established (1952), with the same membership as the earlier Commissions—the Security Council plus Canada. Its task was to prepare proposals for the 'regulation, limitation and balanced reduction of all armed forces and all armaments, for the elimination of all major weapons adaptable to mass destruction, and for effective international control of atomic energy to ensure the prohibition of atomic weapons and the use of atomic energy for peaceful purposes only.' In 1954 the discussions were moved to a subcommittee, consisting of Canada, France, the United Kingdom, the Soviet Union and the United States, where they continued in private until September 1957. A series of disarmament plans consisting of different stages was put forward by each side. By 1955 there appeared to be a considerable degree of convergence. There was agreement, for example, on the eventual force levels, on the total prohibition of nuclear weapons, to be effected after 75 per cent of the reduction of armed forces had been carried out, and on the principle of permanent ground control posts to supervise inspection. However, there was little progress after 1955. In September 1955, the United States representative put a 'reservation' on all earlier disarmament proposals.

Attention shifted to proposals for limited measures such as arrangements for ground and air inspection, the establishment of nuclear-free zone in Europe, measures against surprise attack; and negotiations for the discontinuance of nuclear weapons tests were initiated.

General and complete disarmament

A programme for general and complete disarmament was first put forward by the Soviet Union on 18 September 1959. By 1960 both sides had agreed that general and complete disarmament was the objective of negotiations. The Disarmament Commission was not dissolved but beginning in March 1960 the negotiations were conducted in a new Ten-Nation Disarmament Committee, with five members from NATO countries and five from Warsaw Pact countries, In March 1962 eight non-aligned countries were added to this Committee, thereafter called the Eighteen-Nation Disarmament Committee. To the ENDC the Soviet Union submitted a 'draft treaty on general and complete disarmament under strict international control', and the United States submitted an 'outline of basic provisions of a treaty on general and complete disarmament in a peaceful world'. The Soviet draft treaty provided for the completion of the disarmament process within a fixed, short period of time: nuclear delivery vehicles were to be completely abolished by the end of the first stage of disarmament. The United States outline, on the other hand, provided for gradual disarmament, beginning with a freeze, and keeping the relative military positions throughout the disarmament process similar to what they were at the beginning of the process. The Soviet Union subsequently amended its proposal to permit the United States and Soviet Union to retain, on their territories, a limited number of inter-continental, anti-missile and anti-aircraft missiles until the last stage of disarmament. Little progress was made in these negotiations; and there has not been much discussion of general and complete disarmament since 1965.

2. *International arms control and disarmament agreements: promise, fact and vision*★

The blueprint for disarmament

The decisive turning point in the post World War II disarmament efforts occurred with the US–Soviet 1961 agreement on a program of general and complete disarmament. Though not pursued and consumated, this accord stands out as a sign-post and pointer for contemporary disarmament efforts. It laid down in broad terms the ultimate goals for a disarmed world and designed the general outlines for a systematic and effective disarmament process.

On September 21, 1961, after prolonged discussions, the United States and the Soviet Union presented to the UN General Assembly a Joint Statement of Agreed Principles for Disarmament Negotiations, known popularly as the McCloy–Zorin agreement. This was a document of paramount importance. It contained eight points, the first three of which read:

1. The goal of negotiations is to achieve agreement on a programme which will ensure that (*a*) disarmament is general and complete and war is no longer an instrument for settling international problems, and (*b*) such disarmament is accompanied by the establishment of reliable procedures for the peaceful settlement of disputes and effective arrangements for the maintenance of peace in accordance with the principles of the United Nations Charter.
2. The programme for general and complete disarmament shall ensure that States will have at their disposal only those non-nuclear armaments, forces, facilities, and establishments as are agreed to be necessary to maintain internal order and protect the personal security of citizens; and that States shall support and provide agreed manpower for a United Nations peace force.
3. To this end, the programme for general and complete disarmament shall contain the necessary provisions, with respect to the military establishment of every nation, for:

   (*a*) Disbanding of armed forces, dismantling of military establishments, including bases, cessation of the production of armaments as well as their liquidation or conversion to peaceful uses;

   (*b*) Elimination of all stockpiles of nuclear, chemical, bacteriological, and other weapons of mass destruction and cessation of the production of such weapons;

   (*c*) Elimination of all means of delivery of weapons of mass destruction;

★ Excerpts from a paper by Marek Thee from *International Social Science Journal*, Vol. XXVIII, No. 2, 1976.

(*d*) Abolishment of the organization and institutions designed to organize the military effort of States, cessation of military training, and closing of all military training institutions;

(*e*) Discontinuance of military expenditures.

The other points of the Joint Statement provide for the implementation of the above disarmament program in stages carried out within specific time-limits; for 'strict and effective international control as would provide firm assurance that all parties are honouring their obligations'; for the establishment of an International Disarmament Organization within the framework of the United Nations to execute control over and inspection of disarmament; and for the creation of institutions to maintain world peace and to settle international disputes by peaceful means, including the setting up of UN peace-keeping forces.

Most remarkable in the Joint Statement was the extent of agreement between the two powers. The only disagreement seemed to concern the range of international control and verification. While the United States wanted verification measures to apply to the agreed arms reductions as well as to the retained armed forces and armaments, the Soviet Union was opposed to the establishment of control over armaments which it interpreted as an attempt to introduce a 'system of legalized espionage'.

It is in this context worth-while to note that though problems of control and verification still remain one of the chief stumbling blocks in disarmament negotiations, recent technological advances in remote sensing, seismic detection of underground nuclear explosions and satellite tele-detection alleviated very much the issues. So, for instance, the 1972 Strategic Arms Limitation Talks (SALT) agreements legitimized unilateral 'national technical means of verification', i.e. space-orbiting satellite intelligence gathering as an accepted way of control, and both parties, the United States and the Soviet Union, undertook 'not to interfere' with these verification methods.

Agreements concluded

In the period between 1961 and 1975 a number of international arms control and disarmament agreements were signed and ratified. The existing international arms control and disarmament agreements could be divided into two main categories of a multilateral and bilateral nature, the last concluded between the United States and the Soviet Union. A general classification would show the following subdivisions:

I. Multilateral agreements

1. Accords on nuclear weapons
2. Accords on chemical and bacteriological weapons
3. Accords which tend to exclude armaments in certain world regions or virgin spaces such as the Antarctic, the sea-bed and outer space—so called non-armament agreements.

II. Bilateral US–USSR agreements

1. Accords on measures to reduce the risk of nuclear war
2. Accords on the limitation of strategic arms
3. Accords on nuclear weapon tests.

Following is a table of these agreements with an indication of their main provisions:

I. Multilateral agreements	
1. Nuclear weapons	
1.1 Partial Test Ban Treaty, of August 5, 1963	Prohibits nuclear weapon tests in the atmosphere, in outer space and under water
1.2 Treaty on the Non-Proliferation of Nuclear Weapons, of July 1, 1968	Prohibits the transfer and tends to prevent the proliferation of nuclear weapons.
2. Chemical and bacteriological weapons	
2.1 Geneva Protocol, of June 17, 1925	Prohibits the use in war of asphyxiatic, poisonous and other gases, and of bacteriological methods of warfare
2.2 Convention on the Prohibition of Bacteriological Weapons, of April 10, 1972	Prohibits development, production and stockpiling of bacteriological (biological) and toxin weapons, and provides for their destruction.
3. Regional and virgin space non-armament agreements	
3.1 The Antarctic Treaty, of December 1, 1959	Prohibits any measures of military nature, incl. testing of any types of weapons, in the Antarctic area
3.2 Treaty for the Prohibition of Nuclear Weapons in Latin America (Tlatelolco Treaty), of February 14, 1967	Declares military denuclearization of Latin America (countries which still did not fully acceed to the Treaty include *inter alia* Argentina, Brazil and Chile)
3.3 Outer Space Treaty, of January 27, 1967	Prohibits placing of nuclear arms and other weapons of mass destruction in orbit, on celestial bodies and in outer space
3.4 The Sea-Bed Treaty, of February 11, 1971	Prohibits emplacement of nuclear arms or weapons of mass destruction on the sea-bed, the ocean-floor and in the subsoil thereof.
II. Bilateral US–USSR agreements	
1. Nuclear war risk limitation	
1.1 'Hot-Line' agreement, of June 20, 1963	Establishes direct telegraph-teleprinter communication link for use in emergency

Agreement	Content
1.2 'Hot-Line' Modernization Agreement, of September 30, 1971	Supplements 'Hot-Line' with two additional circuits each using a satellite communication system, and with a multiple system of terminals in each country
1.3 Nuclear Accidents Agreement, of September 30, 1971	Provides for measures to reduce the risk of outbreak of nuclear war, incl. safeguards against accidental use of nuclear weapons
1.4 Nuclear War Prevention Agreement, of June 22, 1973	Provides for restraint and urgent consultations to avert the risk of nuclear war
1.5 High Seas Incident Prevention Agreement, of May 25, 1972	Provides for measures to assure the safety of military navigation and flights on or over high seas.
2. *Strategical Arms Limitation Talks (SALT)*	
2.1 Anti-Ballistic Missile (ABM) Treaty (SALT I), of May 26, 1972	Limits deployment of ABM systems to two sites in each country (to protect the capital and one intercontinental ballistic missiles (ICBM) site)
2.2 ABM Protocol, of July 3, 1974	Limits deployment of ABM systems to one site in each country
2.3 Interim Agreement on the Limitation of Strategic Offensive Missiles (SALT I), of May 26, 1972	Provides for five-year freeze in the aggregate number of fixed land-based ICBMs and submarine-launched missiles (SLBMs) of both parties
2.4 The Vladivostoc accord (SALT II), of November 24, 1974	Provisional agreement on further negotiations to cover the period until 1985; sets a ceiling of 2 400 strategic delivery vehicles in the triad mix of ICBMs, SLBMs and heavy strategic bombers, of which 1 320 may be armed with multiple warheads (MIRVs).
3. *Nuclear weapon tests*	
3.1 Threshold Test ban Agreement, of July 3, 1974	Limits underground nuclear weapon tests to a yield not exceeding 150 kilotons, beginning March 31, 1976 peaceful nuclear explosions excluded.

Value and utility

What is the value of these agreements? What is their significance and utility?

No doubt, in historical terms, the arms control and disarmament agreements signed in recent years mark a new departure in disarmament efforts. Never in history were so many arms control agreements concluded in such a short period. This reflects the awareness of the growing dangers to humanity generated by the arms race, a race intertwined today with the second technological revolution and advanced stages of the nuclear era. Whatever the deficiencies of the concrete disarmament measures inscribed in the respective accords, as a whole they contribute to some relaxation of international tension. One might then hope that the disarmament negotiations and debates may serve as a learning, educational and corrective process in international relations, in which suspicions are extenuated and mutual confidence promoted, rigidities are blunted and restraint stimulated, Cumulatively, one would like to believe that technological strides in the development of new arms will be matched by political will and resolve of the powers to actively seek real disarmament.

At the same time we have to be conscious and sensitive of the shortcomings and limitations. For on their overcoming depends genuine success of arms control and disarmament.

Different criteria can be used to judge the value and utility as well as the shortcomings of the agreements concluded. One possibility is to measure their actual provisions and achievements on the scale as envisaged by the program of general and complete disarmament. Another way of evaluation is to apply the accords to the overall world or national security balance or imbalance, i.e. to try to find out how far and if international security has been enhanced, considering especially the parallel accumulation of new weapons. Still another method of assessment is to compare the levels of armaments and military expenditures before and after the conclusion of the agreements. Other criteria could yet be conceived such as the changed dynamics of the arms race, the interrelation between negotiations and the armaments course, the virtue of the very negotiations process or, last not least, historical lessons from disarmament negotiations. Perhaps some elements of all of these approaches are needed to fully appraise the agreements reached.

To start with, we need first to differentiate between the concepts of disarmament and arms control. Disarmament, as the very term suggests, means tangible elimination or reduction of armaments in real quantities, implemented on the basis of international agreements and aiming for lasting solutions in preserving peace and security. There may also be unilateral disarmament moves undertaken, for instance, in the hope of reciprocity and generation of a chain reaction of disarmament measures. Arms control, on the other hand, has been defined as measures aiming (*a*) to reduce the probability of war, (*b*) to reduce the costs of preparation for war, and (*c*) to reduce the death and destruction if control fails and war comes. In other words though arms control may lead to disarmament measures, it does not necessarily assume reduction of military forces or armaments. Its focus is, actually, on the control of the process of armaments so as to keep a balance between the military capabilities of the main adversaries, and to avoid a destabilization of the international military environment. It is concerned with limiting the risk of war breaking out in situations of crisis, and with reducing human and material losses if hostilities nevertheless erupt. Real disarmament, in the view of arms controllers, is a problem of the distant future and may only come as a result of a series of partial and selective confidence-building arms control measures which would pave the way for a change of political postures.

Examining the list of armaments accords signed in recent years, we immediately notice that almost all of them fall within the category of arms control agreements. Only one of the accords, the 'Convention on the Prohibition of the Development and Stockpiling of Bacteriological (Biological) and Toxin Weapons, and of their Destruction' provides for the elimination of a concrete weapon, though it does not envisage international control and verification of the destruction of these weapons. It is also striking that none of the agreements deals with traditional conventional weapons as used in World War I and II, and still the main tool of death and destruction in so many local conflicts after World War II.

It becomes thus clear that in the process of disarmament negotiations of the last decade or more, the idea of general and complete disarmament, of comprehensive disarmament measures, has been discarded, left only alive in declarations of intent in preambles of some of the accords concluded. The notion of disarmament was reduced to limited partial arms control measures focussing on the ABC weapons—atomic, biological and chemical.

From this point of view, the agreements concluded may fall into the following categories:

1. Accords of a declaratory political nature pronouncing restraint in military action to avoid the risk of nuclear war (II.1).
2. Non-armament agreements aiming to keep free of nuclear weapons or weapons of mass destruction certain geographical regions or outer space and sea-bed environments, of lesser immediate interest for the actual arms race (I.3).
3. Accords on partial steps tending to limit the horizontal spread of nuclear weapons (I.1 and II.3).
4. Accords aiming to put a quantitative ceiling on certain strategical arms, defensive and offensive, crucial for nuclear warfare (II.2).
5. Accords concerning chemical and bacteriological weapons, with the 1925 Geneva no-first-use Protocol still the main achievement (I.2). Note: no such treaty commitment exists in relation to nuclear weapons.

Scope and meaning

Important as the above agreements might be, they have one basic flaw: they did not halt the arms race. In the last fifteen years, while these agreements were negotiated and concluded, world armaments expenditures have doubled in constant prices, approaching to-day the staggering figure of

300 billion dollars annually. Nuclear weapon testing, in keeping with the 1963 Partial Test Ban Treaty, has only changed environment moving from tests in the atmosphere to tests underground. Moreover, according to available data, testing became more intense and frequent. From 1960 the number of powers which carried out nuclear explosions has doubled adding France, China and India to the members of the nuclear club. About a dozen other states, the so called near-nuclear or threshold nations, are known to have acquired nuclear capabilities though they did not resolve upon nuclear testing. Both conventional and nuclear arms, revolutionized by new technology, acquired unprecedented levels of sophistication and destruction. The volume of international arms trade has increased several times with most modern conventional weapons flowing to-day to all corners of the world. The world arsenals of nuclear weapons and delivery vehicles grew steadily reaching an overkill capability sufficient to destroy mankind several times over.

Promising as they were in the beginning, the partial arms control measures did not reverse trends and did not lead to comprehensive disarmament measures. In reality, the partial approach channelled all efforts to accords in fields of limited or marginal value from the point of view of armaments, leaving almost untouched or loopholes in spheres of primary interest for the military. Thus, nuclear weapon testing was banned in the atmosphere, an important environmental measure, but was allowed to continue underground. And the 1974 Threshold Test Ban Agreement, permitting tests of a size ten times greater than the Hiroshima bomb, approved a threshold high enough to allow further weapon development. Repeated demands for a comprehensive nuclear test ban, as advanced in the Geneva Conference of the Committee for Disarmament and in many UN General Assembly resolutions, were left unheeded. Similarly, the Nuclear Non-Proliferation Treaty prohibited the transfer of nuclear weapons to non-nuclear-weapon states but did not restrict perfection of these weapons by the nuclear powers. Emplacement of fixed nuclear devices on the sea-bed and the ocean-floor was banned, but the military nuclear fleet including nuclear submarines could continue to roam the world seas and oceans. Biological weapons which were never used had to be destroyed but chemical weapons which were used and pose a much greater threat to humanity were not covered by the convention. Efforts were made to set a quantitative ceiling to strategical arms, which anyhow reached a great overkill capability, while qualitative improvements of those arms impelled by a huge military research and development establishment could freely proceed, expressly permitted by the SALT agreements (Art. VII of the ABM Treaty and Art. IV of the Interim Agreement). The actual effect of the agreements reached was thus to lower the visibility of the arms race and limit armaments in spheres of secondary military import, while at the same time legalizing and rationalizing arms development in fields of actual military utility.

There is a crucial time factor to these developments. The disarmament negotiations do not keep pace with the arms race, nor does arms control cope with the technological arms development dynamics. The arms race and the technological armaments dynamics by far outstrip the efforts of the arms controllers. Of decisive importance is the fact that armaments in recent years were channelled from a race in quantities to a race in qualities, upheld by almost half a million of scientists and engineers employed by military research and development. Whenever arms control negotatiations are approaching an agreement to balance existing types of weapons, new kinds of arms threaten to destabilize the situation. In fact, prolonged negotiations tend to feed efforts to invent and produce new weapons, first as a 'bargaining chip' in the talks but in the process assimilated by the military. This was best observed in the protracted SALT negotiations which from stage to stage had to deal with the deployment of ever new weapon systems such as the ABMs, MIRVs, cruise missiles or new types of bombers. In times of rapid technological advances there seems to be no substitute for comprehensive measures if disarmament is to produce real results.

Arms controls as practiced in the last 10–15 years has thus become a disappointing exercise. The more we analyze the arms control negotiations, the agreements reached and the dynamics of the arms race, we come to realize that the situation in armaments is not improving but deteriorating. The reorientation from disarmament to arms control and partial measures was perhaps initially induced by the complexities of the tasks but, in the process, arms control became an exercise tailored to the needs of the two dominant powers, the United States and the Soviet Union, to maintain a world strategical balance. The outcome was differently appraised by different people. Some felt content that nuclear war was prevented. Others were disturbed that the arms race is continuing. It seemed a vicious circle dominated by uncertainties, in which armaments were praised and charged for maintainance of peace on the one hand, and its undermining on the other—of having a stabilizing and destabilizing effect at the same time.

But outside the two power blocks another aspect came to cause serious concern: the power balancing act, aside from the constant perils of destabilization, contributed to maintain and preserve a duopoly in high technology and a general world military and economic predominance of the two powers. And despite some appearances, the balance of power game did not enhance either security or stability. The more arms around and the more they are refined, the more delicate the balance and greater the chances of destabilization. The upward spiral of the arms race with ever larger and larger nuclear arsenals and more sophisticated delivery vehicles increases only the vulnerability of the adverse parties and the capability to destroy each other, makes first strike options more tempting and second strike retaliation less fearful or even probable. Thus, the deterrence rationale, fallacious in itself, tends to break down. Moreover, the apparent deterrence stability as conceived by some from the perspective of bipolar superpower relations, may crumble altogether if the deterrence equation is broadened to include a number of third, actual and aspiring, nuclear states. There is broad consensus to-day among disarmers and arms controllers that we may land in a world more dangerous than ever if nuclear armaments are not halted in time.

Concluding remarks

A survey of the arms control and disarmament agreements in force, negotiated and concluded in the last fifteen years, leaves us with a feeling of acute insufficiency and inadequacy. In fact, nobody can be satisfied with the actual state of affairs. Concern has often been expressed to this effect at the highest levels in all capitals of the world, all the great powers included. While Soviet Foreign Minister Andrei Gromyko pointed to the incompatibility between the aim of a stable and lasting peace, and the on-going arms race, US Secretary of State Henry Kissinger warned that 'nuclear catastrophe looms (to-day) more plausible, whether through design or miscalculation, accident, theft or blackmail'. We may sometimes feel relieved that a process of arms control negotiations has been initiated, and can also be glad with each separate accord. Yet at the same time, we are painfully aware of the basic reality that arms control negotiations do not catch up with the pace of armaments and technological innovation. Despite the negotiations and the agreements signed, the arms race continues unabated.

Another material element of the actual arms control accords also deserves attention. It is their temporary character, very much dependent on the political climate. Almost all of the agreements concluded in recent years, from the Partial Test Ban Treaty and the Nuclear Non-Proliferation Treaty, to the Biological Convention and the SALT agreements, contain the crucial clause that 'each Party shall in exercising its national sovereignty have the right to withdraw from the Treaty if it decides that extraordinary events, related to the subject matter of this Treaty, have jeopardized the supreme interests of its country'. Renunciation of treaties when powers feel it convenient for them, as we know, is a frequent occurrence in history. There is then a need to reflect on such withdrawal clauses and think of solutions which would make disarmament irreversible. Again, it seems that this may be possible only within a framework of comprehensive measures.

Furthermore, essential for the success of disarmament efforts is a global perspective and the universality of accords. With the advance of the nuclear era (some call it the second nuclear era), the rapid proliferation of nuclear technology, and the development of intercontinental ballistic and non-ballistic precise delivery vehicles, the concept of collective security acquires new vitality and meaning. Fine-tuning of the arms race between two powers or blocks, in itself sometimes like ploughing the sands, may prove futile if the rest of the world, other powers and nations, are not part of the peace-keeping effort. At present, two nuclear powers, China and France, do not participate in any disarmament negotiations, and there are also many nuclear have-nots whose participation may be essential. In fact, we approach a time when world security will depend to an ever larger degree on the cooperation of the whole family of nations, on the pooling of forces and efforts.

In problems of armaments and disarmament mankind stands to-day on momentous crossroads. Our time is fraught with peril as never before. Not only are we threatened by stockpiles of terrible arms capable to inflict irreparable damage to men and their environment, but we have to face also a dynamics of armaments with unpredictable consequences. A co-ordinated effort of a small group of scientists during World War II produced the atom bomb. Today, on a much higher level of scientific knowledge, hundreds of thousands of scientists and engineers serving the military are competing across nations and within nations to develop and master new weapons. This has revolutionized military science and technology, and cannot but produce spectacular results.

The conclusions to draw from this kind of development are crucial for present-day arms control and disarmament negotiations. As the center of gravity of armaments moved from quantity to quality, no arms control or disarmament agreement can possibly produce lasting positive results without addressing ourselves to the problem of military research and development, its scope, mode of operation and performance. The limited momentum in arms control achieved in the post World War II period is put to a most serious trial.

*3. The game of disarmament**

'Are the two nuclear super-powers (United States, USSR) really serious about nuclear disarmament?'

The following article is an attempt to answer this harrowing question in a personal way, from the vantage point of a participant in ten years of disarmament negotiations, where I have been representing one of the non-aligned, non-nuclear-weapons states.

Let me immediately reveal how the experiences of a decade of disarmament efforts have gradually led me and many others participating in these efforts to some terrifying conclusions: that we have accomplished no real disarmament, that we can see hardly any tangible results of our work, and that the underlying major cause must be that the superpowers have not seriously tried to achieve disarmament.

As matters stand today, my severe judgement about the lack of achievement in disarmament negotiations can be fully substantiated by a simple reference to the steep increase in actual armaments which has taken place while the talks on disarmament have gone on.

All nations seem to be bent on continuing down the road to more abundant and more lethal armaments. But the primary actors are without doubt the nuclear-weapons superpowers. Their overkill capacity continues to spiral upwards. The number of nuclear warheads as well as of their delivery vehicles has increased in an alarming manner since the start of the

* Shortened version of a paper by Alva Myrdal from *Impact of Science on Society*, Vol. XXII, No. 3, 1972.

disarmament negotiations in Geneva ten years ago. No less significant is the increase in accuracy and destructive power of all weapon types, the automation of systems of targeting and communications, the breakthroughs in military electronic equipment, the vastly improved cameras for spy satellites, etc.

The case of the test ban

Disarmament has not failed for lack of effort. When looking back at the decade of internationally institutionalized and remarkably sustained disarmament negotiations, one is struck by the high ambitions held at the start in March 1962 of the ENDC (Eighteen Nation Disarmament Committee, as it was then called). Both the United States and the USSR each submitted a draft Treaty on General and Complete Disarmament.

The initiative was actively seized in the very first week by the group of newcomers: the non-aligned countries which had just been added to those of the two blocks which had 'off-and-on' been entering into disarmament talks practically since the end of the Second World War. Within a month the non-aligned members presented a memorandum (April 1962) offering a new scheme for a comprehensive ban on testing of nuclear weapons.

A total ban on testing of nuclear weapons would have erected a double barrier of formidable strength: it would have meant the end of the qualitative development of nuclear weapons, thus the end of the competitive race between the nuclear powers, and, in addition an effective block—for all practical purposes—against any acquisition of such weapons on the part of non-nuclear-weapons countries which might be tempted to veer away from the narrow path of self-denial staked out for them.

Soon we could witness the first side-stepping of this crucial issue—although we have until this day hesitated to acknowledge what happened for what in reality it was. The accent was shifted from stopping tests as a means to stop further development of nuclear weapons, to stopping them instead in order to avoid the side effects of atomic radiation.

That the existence of the nuclear weapons themselves was the larger worry was somehow lost sight of. Anyway, pressure within the United Nations was not enough to get the superpowers to bow down and cease the development, testing and production of nuclear weapons.

I must confess that we did not wake up to understand the sombre reality even when the superpowers suddenly, in the summer of 1963, switched the test-ban negotiations from Geneva, where we were in good faith continuing to labour on a *total* test ban, to bilateral talks in Moscow, where within weeks was produced a *partial* ban. So doped in hope were we that we euphorically hailed this agreement as one of utmost importance. We took it for granted, as we were told, that it was the first step towards the discontinuance of all tests.

The Preamble of the Moscow Treaty (Partial Test Ban Treaty) explicitly spelt out the commitment of the parties as 'seeking to achieve the discontinuance of all test explosions of nuclear weapons for all time, determined to continue negotiations to this end'. We read the intention to be that although the Moscow treaty provisionally excluded underground tests from the partial prohibition, it would nevertheless serve as a disarmament measure by curtailing further qualitative development of nuclear weapons, to a degree which was undetermined but understood to be not insignificant.

It has since become irrefutably clear that the truth is different: the Moscow treaty has not had any restrictive effect whatsoever on nuclear-weapons development or even on the number and yield of tests made by those nations who already possess such weapons. It should be given some credit as a public-health measure, since it has reduced radio-activity in the atmosphere—even if the degree of such pollution had never been high—but it can hardly any longer correctly be enumerated among disarmament measures.

Our naïveté and credulity were such that we smaller nations did not at the time realize that no disarmament was intended. We evidently did not listen closely enough, for the absence of this intention was clearly stated in President Kennedy's own words when introducing the Moscow treaty to the United States Senate. He then explicitly promised the military interests that the country's testing facilities would not be closed: 'The United States has more experience in underground testing than any other nation, and we intend to use this capacity to maintain the adequacy of our arsenal. Our atomic laboratories will maintain an active development program, including underground testing, and we will be ready to resume testing in the atmosphere if necessary.'

With the hindsight of today the conclusion of the Moscow treaty should perhaps be judged as a political mistake. In fact, it introduced the practice of sealing off disarmament schemes with a full stop, as soon as some token measure of success could be registered and sold to the general public—however partial, one-sided or illusory.

The Non-Proliferation Treaty

To understand both what has happened and what has not happened as far as disarmament is concerned, an analysis must seek to discern what interests have been at play. *Cui bono* is the question that provides the key to the understanding of the 'game of disarmament'.

Let us concentrate on the nuclear-weapons field and explore what has really happened behind what seems to have happened in relation to the supposedly second most important achievement of disarmament negotiations. I refer to the Non-Proliferation Treaty (NPT), which is primarily designed to prevent an increase in the number of countries possessing nuclear weapons.

This is a worthwhile purpose *per se*. But everybody knows that the greatest dangers spring not from the non-nuclear-weapons States. Eager as we, the latter, are to promote nuclear disarmament among the countries which do possess such weapons, we raised the demand that a ban on 'horizontal proliferation' (spread of nuclear weapons to more nations) should be combined with one on 'vertical proliferation' (further production of such weapons by States already possessing them). This would mean establishing a ceiling on the total nuclear-weapon strength in the world at the level pertaining at the time of concluding the treaty.

In order to make such a freeze on nuclear-weapons development complete, Sweden for several years advocated that the package deal should include as a third element a comprehensive test ban, thus freezing qualitative as well as quantitative proliferation of nuclear weapons everywhere.

On this issue the power position of the superpowers was asserted even more bluntly than in the case of the Moscow treaty. In the case of the Non-Proliferation Treaty they accepted not one iota of sacrifice of present or future nuclear-weapons capabilities. The only trace of the years of negotiating effort expended by the smaller countries can be found in an article in the NPT which enjoins the contracting parties 'to pursue negotiations, in good faith, on effective measures relating to cessation of the nuclear arms race at an early date'.

The harsh reality

The above reflections on the hopes and disappointments of a non-aligned conscientious collaborator in disarmament striving, when judging the outcomes of two of the major so-called disarmament agreements, can but lead to a bitter verdict. This is that disarmament has for the great powers become an object of super-salesmanship, whose real objective is to get the smaller nations to accept restrictions while—at least up to now—not being willing to accept any themselves.

This leads us to ponder over the very 'whys' of international treaties in such a field as nuclear weapons. Why make them international–multi-national?

The question as to why disarmament treaties are negotiated multi-nationally can be put even more pointedly in regard to those agreements which are not only arranged by bilateral superpower accord, such as those just mentioned, but which are in substance relevant only to the nuclear-weapons States. Such is the case in regard to another pair of disarmament agreements, the 1970 treaty concerned with keeping the sea-bed under international waters free of nuclear weapons and the Space treaty of 1967, which similarly safeguards outer space and bodies in space. Both have in common the fact that only the superpowers have practical access to these domains and only the nuclear-weapons countries could, in any case, have any forbidden objects to place there.

The case of the Space treaty, on the peaceful uses of outer space, is particularly interesting in this respect. The real agreement, which is *de facto* operational, was a bilateral one announced to the United Nations in October 1963 by Kennedy and Gromyko. Its transformation in 1967 into an international treaty with a number of signatories who cannot, however much they might wish, place nuclear weapons on celestial bodies or in orbit must surely look pointless when this transformation is assessed for its disarmament effect.

I want to underline that my criticism is not directed against bilateral negotiations—their value will be proved by their results—but against the instituting of multilateral negotiations if they are allowed to be no more than a shadow play, or, as the Ethiopian delegate in Geneva recently said, 'a window-dressing'.

Any attempt at providing my personal and honest response to the question posed as the topic for this article must result in a negative answer, on balance, when weighing the pious talk about disarmament against the harsh reality of the arms race. It becomes unavoidable to conclude that underlying this answer is the lack of any real effort on the part of the superpowers.

Hegemony over world affairs

In an effort to explain the negotiation behaviour of the superpowers in regard to disarmament, I have already indicated that the reason behind their desire to present some 'positive results' in the shape of international conventions, even when of marginal real value, has been the wish to make some concessions to the clamour of world public opinion. In bleak moments one might think that they just want to keep the would-be disarmers busy: negotiations as a kind of occupational therapy.

The reasons for the 'negative results'—i.e. the lack of readiness to achieve more—must be of a very much more complex nature.

I have already repeatedly stated how obviously the disarmament negotiations have been handled by the superpowers as in substance bilateral, even when done within an international framework. This preoccupation of theirs with each other rather than with the world has led them to use two different tactics, depending on whether the game is played between the two of them alone or by both of them against the other nations.

In regard to each other the two superpowers have landed in an arms race which somehow resembles a zero-sum-game: the loss of one is considered the gain of the other. Neither dares to give up a military capability or even an option for a new weapons system lest he may be 'losing ground to the enemy'. The consequence has been that the disarmament negotiations have been used by the superpowers for *balancing each other* and not for *planning disarmament*.

Only gradually has it dawned upon us—and not least upon me—that the arms race, however 'balanced', was having as its result an incredible widening of the gap between the superpowers and the rest of the world. This split into two discontinuous categories of 'superpowers' and 'other nations' has not only become more apparent to us during the disarmament negotiations, it has been made even more bluntly manifest by a conscious design on their part. The best example of this is, of course, the NPT. The NPT in no way infringed upon the nuclear-weapons powers' freedom to develop, test, produce, store, deploy or use these weapons.

The mounting discrimination against the smaller nations is, however, not only of concern in regard to the military monopolization of ultimate weapons on the part of the superpowers. It also has considerable spin-off effects in other spheres. Monopoly over military technology is beginning to play a cardinal role in a similar monopoly over new technologies of immense importance in the economic life of nations and in the relations between powers in general. What we are witnessing today, it seems to me, is the emergence of a duopoly of the two superpowers in regard to modern technology, giving them a more and more dominating hegemony over world affairs.

In this article, I have considered the negative or at least tokenistic, attitude towards true multilateral disarmament on the part of the superpowers as being determined by policy considerations. I stand by the main conclusions of this analysis, even if they sound like a *j'accuse*.

Will reason prevail?

Is it not possible to end on a more optimistic note?

There always remains the chance that reason may begin to prevail. It may come to dawn on the policy-makers of the world's States that the tremendous waste of resources—financial, material, technological or human, in bodies and brains—on purposes of destruction might be turned into tremendous progress for the benefit of peoples all over the world.

A very special hope which I cannot help nourishing is that scientists, not least those now working for the war machines or on peace research, will begin to oppose more forcibly the present acceptance of the rule of unreason—so forcibly that they will be heard.

*4. The implementation of the nuclear Non-Proliferation Treaty**

On 5–30 May 1975, five years after the entry into force of the Treaty on the Non-Proliferation of Nuclear Weapons (NPT), a conference was held in Geneva to review the operation of the treaty. Of a total of 96 states party to the NPT at the time of the Conference, only 58 or 60 per cent attended.

The Review Conference concluded its work with the adoption, by consensus (without a vote being taken), of a Final Declaration. However, in spite of the formal acceptance of the declaration, a number of delegations expressed dissatisfaction about the outcome of the conference, made interpretative statements contradicting the consensus, or objected outright to

* From *SIPRI Yearbook*, 1976.

various formulations. Proposals for additional protocols to the NPT, as well as resolutions dealing with various matters related to the implementation of the NPT, were submitted by several participants but did not obtain sufficient support. On the insistence of the sponsors, they were included in the Final Document of the Conference for subsequent consideration by the governments of states party to the NPT.

Non-transfer and non-acquisition of nuclear weapons

The first two articles of the NPT contain the essence of the non-proliferation undertakings. Under Article I the nuclear-weapon states are committed not to transfer, while under Article II the non-nuclear-weapon states are under the obligation not to receive, manufacture or otherwise acquire, nuclear weapons or other nuclear explosive devices, or control over them.

The Review Conference declaration contends that these articles have been observed by all parties. Nevertheless, it would be wrong to conclude that the very purpose of the NPT has been achieved. Since the treaty has not been universally subscribed to, its observance by the parties alone cannot guarantee a halt to nuclear-weapon proliferation. In fact, the number of states known to possess nuclear weapons or other nuclear explosive devices, which the NPT was intended to restrict to five, increased when India carried out a nuclear explosion. But India cannot be charged with a breach of the NPT which it never signed. Even before the NPT was signed, India and a few other near-nuclear-weapon states had made it clear that they would not adhere to it.

No sooner had the political fallout from the Indian nuclear explosion settled, when a new, even more disturbing act related to the NPT was revealed. As a result of secret talks started in 1974, an agreement was signed on June 27 1975 between the Federal Republic of Germany and Brazil 'on cooperation in the field of peaceful uses of nuclear energy'. Never before has such a comprehensive nuclear deal been concluded. It is also conspicuous by its sheer size: an estimated DM 12 billion is to be spent in the Federal Republic of Germany over the next 15 years. FR Germany, in addition to immediate commercial gains, may get ensured access to the deposits of Brazilian uranium. The deal signifies the creation of a new self-sufficient nuclear state with a nuclear-weapon capability.

The Federal Republic of Germany declared that it would share its know-how and experience in the peaceful uses of nuclear energy only with countries which have decided to renounce the manufacture or acquisition of nuclear explosive devices. But Brazil does not belong to this category of states: it has refused to join the NPT; and although it has signed and ratified the Treaty for the Prohibition of Nuclear Weapons in Latin America (Treaty of Tlatelolco), it has not waived the requirements for the entry into force of the treaty, as laid down in Article 28, and is therefore not bound by its provisions. Brazil (as well as Argentina) insists on the right to carry out nuclear explosions for peaceful purposes, 'including explosions which involve devices similar to those used in nuclear weapons', that is, the right to do something that is explicitly forbidden by the NPT and implicitly prohibited by the Treaty of Tlatelolco as interpreted by other parties to the treaty.

The Review Conference rightly pointed out that strict observance of Articles I and II is 'central to the shared objective of averting the further proliferation of nuclear weapons'. However, it is precisely the failure to observe strictly these articles, in particular the obligation not to assist 'in any way' a non-nuclear-weapon state to manufacture nuclear explosive devices, that contributes to the weakening of the NPT, thus raising the question as to what extent non-proliferation really is a 'shared' objective.

Nuclear safeguards

Under Article III of the NPT, the non-nuclear-weapon states undertook to conclude safeguards agreements with the IAEA covering all their peaceful nuclear activities. Although the control provisions constitute an inseparable part of the NPT commitments, not all non-nuclear-weapon parties have concluded the required agreements. The NPT Review Conference recognized that this was an unsatisfactory state of affairs by emphasizing the 'necessity for the States Party to the Treaty that have not yet done so to conclude as soon as possible safeguards agreements with the IAEA'. No definitive date has been set.

The Conference declaration attached considerable importance to the continued application of safeguards to the nuclear activities of the non-nuclear-weapon parties to the NPT, 'on a non-discriminatory basis'. It failed, however, to settle the problem of discriminatory treatment of the parties to the NPT as compared with non-parties. As a result of the policies of the suppliers, the latter, as distinct from the former, are not subject to safeguards comprehensively covering their nuclear activities; safeguards applied in their territories are facility-oriented, which means that they may put nuclear material only in certain facilities under IAEA safeguards and retain unsafeguarded all or part of a nuclear fuel cycle. There can, therefore, be no guarantee that non-peaceful nuclear activities are not carried out on the territory of the recipient states, non-party to the NPT. As pointed out by the IAEA Director-General during the NPT Review Conference, there will be no overall satisfactory safeguards system operating 'until suppliers of equipment and materials make it a condition for delivery that the entire nuclear activity in the receiving country is placed under the IAEA safeguards'.

The preamble to the Final Declaration of the Conference stressed that the absence of 'effective safeguards' will under the conditions of the accelerated spread and development of peaceful applications of nuclear energy contribute to further proliferation of nuclear explosive capability, but it failed to draw all conclusions from this important statement.

Peaceful uses of nuclear energy

Article IV of the NPT deals with the contribution by states, in a position to do so, to the development of the applications of nuclear energy for peaceful purposes, 'especially in the territories of non-nuclear-weapon States Party to the Treaty, with due consideration for the needs of the developing areas of the world'.

The implementation of this article was seriously questioned by many participants at the Review Conference when statistics showed that non-parties to the NPT had benefited considerably more from international exchange in the field of the peaceful uses of nuclear energy than had the parties to the treaty. Attempts at establishing common standards were made in 1975 by Canada, FR Germany, France, Japan, the UK, the USA and the USSR, the industrial nations responsible for most nuclear material supplies on the world market. In November 1975, this so-called group of seven, meeting in London, reached a gentlemen's agreement (subsequently formalized by an exchange of letters) laying down guidelines for nuclear transfers to any non-nuclear-weapon states for peaceful purposes, and setting certain common safeguards requirements.

The guidelines for nuclear transfers provide for consultations among the seven on matters related to their implementation, presumably before concluding a major export contract, but they contain nothing that would indicate a determination to refrain from selling to individual countries such critical equipment as uranium enrichment and plutonium reprocessing facilities. Notwithstanding certain innovations, non-parties to the NPT will continue to be subject to less comprehensive safeguards than the parties, and adherence to the NPT will still not entitle the latter to favoured treatment in nuclear supplies.

Admitting that wide availability of nuclear technology and fissionable material aggravates the possibility of nuclear-weapon proliferation, the Review Conference considered a proposal for setting up regional or multinational nuclear fuel cycle centres. Large centralized facilities could reduce national incentives to develop individual enrichment and reprocessing plants (enrichment and reprocessing are uneconomical unless tied to a large number of power reactors), satisfy the need for an assured supply of fuel and make technical assistance to less developed countries more effective. Moreover, the application of safeguards would be facilitated and physical protection of nuclear material enhanced due to a decrease of transportation risks. The idea of multinational fuel cycle centres was considered within the group of seven supplier nations, but the discussion was inconclusive.

Peaceful nuclear explosions

Article V of the NPT contains an obligation to ensure that potential benefits from any peaceful applications of nuclear explosions should be made available to non-nuclear-weapon states party to the treaty.

The Conference Declaration noted that the technology of nuclear explosions for peaceful purposes was still 'at the stage of development and study' and that there were a number of interrelated aspects of such explosions which still needed to be investigated. Sweden suggested that if the studies showed that, on balance, no considerable benefits could be derived from peaceful nuclear explosions, states should refrain from their use. The Western countries were very sceptical as to the usefulness and feasibility of these explosions, and the USA seemed to favour the idea of giving them up altogether. (US experiments have been disappointing and no new field tests are currently scheduled.) However, most other countries, including the USSR which claims some achievements in the field of peaceful nuclear explosions, continued to believe in the economic gains they may produce, and that environmental hazards could be avoided.

Some countries suggested a suspension of peaceful nuclear explosions until their economic value had been proved and the problem of their compatibility with a comprehensive test ban solved. It would seem, however, that a temporary measure could be meaningful only if a moratorium on all nuclear tests, both military and non-military, were declared, as there is a direct relationship between these two types of testing.

Disarmament obligations

Article VI of the NPT contains a commitment to pursue negotiations 'in good faith' on effective measures relating to the cessation of the nuclear arms race at an early date and to nuclear disarmament, as well as on a treaty on general and complete disarmament under strict and effective international control.

The non-nuclear-weapon participants at the Review Conference, in particular representatives of the nonaligned countries, drew attention to and showed concern about the continuing nuclear-weapon test programmes and the steady increase of nuclear arsenals in spite of the negotiations on their limitation. In response to the Soviet contention that the basic problems of nuclear disarmament can only be solved with the participation of *all* nuclear powers, opinion was expressed that the USA and the USSR, being the most powerful nations, should take the lead in the disarmament process, thereby encouraging other states to join.

A group of 20 states—Bolivia, Ecuador, Ghana, Honduras, Jamaica, Lebanon, Liberia, Mexico, Morocco, Nepal, Nicaragua, Nigeria, Peru, the Philippines, Romania, Senegal, Sudan, Syria, Yugoslavia and Zaire—suggested the adoption of a protocol to the NPT (Additional Protocol I), under which the depositary governments—the UK, the USA and the USSR—would undertake 'to decree the suspension of all their underground nuclear weapon tests for a period of ten years', as soon as the number of parties to the NPT reached 100; to extend by three years the moratorium contemplated above, each time that five additional states became party to the NPT; and to transform this moratorium into a permanent cessation of all nuclear-weapon tests through the conclusion of a multilateral treaty for that purpose, as soon as the other nuclear-weapon states indicated their willingness to become party to it. The sponsors of the protocol expressed the conviction that the proposed document would in no way undermine the security of the depositary states since the extent of the lead in nuclear-weapon technology and the enormity of the nuclear arsenals of the USSR and the USA were such that 'even if they were to suspend all nuclear weapon tests for half a century, it is absolutely certain that they would continue to maintain an indisputable superiority'.

The same group of nations, with the exception of the Philippines, also proposed the acceptance of another protocol (Additional Protocol II), by which the USA and the USSR would undertake, as soon as the number of parties to the treaty had reached 100, to reduce by 50 per cent the ceiling of 2 400 nuclear strategic delivery vehicles contemplated for each side under the 1974 Vladivostok accords; and reduce likewise by 50 per cent the sub-ceiling of 1 320 strategic ballistic missiles which, under those accords, each side may equip with multiple independently targetable re-entry vehicles (MIRVs). The governments concerned would also undertake, once such reductions had been carried out, to reduce by 10 per cent the ceilings of 1 200 strategic nuclear delivery vehicles and of 660 strategic ballistic missiles that may be equipped with MIRVs, each time that 10 additional states became parties to the NPT. Other clauses would be the same or similar to those of the first protocol. The sponsors argued that the extent of the lead of the USA and the USSR was such that 'even after they had carried out the parity reductions called for in the Additional Protocol, the number of nuclear weapons and of delivery vehicles which each one would maintain would still be much superior to that which might be at the disposal of all the other nuclear-weapon States taken together'.

The above proposals proved completely unacceptable to the nuclear-weapon states. They refused to discuss any time-table for nuclear arms-control measures, even though, according to the NPT, such measures should be carried out 'at an early date'. They contended that the Review Conference was not competent to deal with a matter which was their exclusive concern. The Soviet Union qualified these attempts as inadmissible interference with US–Soviet relations and the USA stressed that it was up to the SALT negotiators to determine the pace of progress in nuclear arms limitation.

The preamble to the Conference Declaration recognizes that it is essential to maintain in the implementation of the NPT an acceptable balance of mutual responsibilities and obligations of all the parties to the treaty. The additional protocols were presented with a view to redressing the balance by matching the cessation of 'horizontal' proliferation with a halt to 'vertical' proliferation. A treaty denying a powerful weapon to most nations in order to preserve a firebreak between the 'haves' and 'have-nots' is not likely to withstand the pressures of a continued arms race. Since nuclear weapons appear to have political and military usefulness for the nuclear powers, the non-nuclear-weapon countries may feel that they too must obtain these advantages. A dynamic process of nuclear disarmament is therefore necessary to de-emphasize the role of nuclear weaponry in world diplomacy and military strategy and to generate political and moral inhibitions dampening the nuclear ambitions of non-nuclear-weapon states.

The Final Declaration contains an appeal to the nuclear-weapon parties to the NPT to make every effort to reach agreement on the conclusion of an effective comprehensive test ban. It notes that a considerable number of delegations expressed the desire that the nuclear-weapon states party to the treaty should as soon as possible enter into an agreement, open to all states and containing appropriate provisions to ensure its effectiveness, 'to halt all nuclear weapons tests of adhering states for a specified time, whereupon the terms of such an agreement would be reviewed in the light of the opportunity, at that time, to achieve a universal and permanent cessation of all nuclear weapons tests'.

Furthermore, the Conference appealed to the USA and the USSR to endeavour to conclude at the earliest possible date the negotiations on 'further limitations of, and significant reductions in, their nuclear weapons systems'.

On the initiative of Romania, the United Nations was invited to consider ways and means of improving its existing facilities for the collection, compilation and dissemination of information on disarmament issues, 'in order to keep all governments as well as world public opinion properly informed on progress achieved' in the realization of the provisions of Article VI of the NPT. The USSR did not support this proposal. In its view the existing organs of the United Nations 'suffice to ensure that all states and world opinion are informed on such issues'.

The security of non-nuclear-weapon states

In a UN Security Council resolution adopted on June 19 1968, the states renouncing the acquisition of nuclear weapons under the NPT had already received a pledge of immediate assistance, in accordance with the UN Charter, in the event they became 'a victim of an act or an object of a threat of aggression in which nuclear weapons are used'. (The nature of the assistance was not specified.) But the value of this document has been repeatedly questioned on the following grounds: *first*, the resolution and the declarations by the UK, the USA and the USSR, associated with it, merely reaffirm the existing UN Charter obligation to provide or support assistance to a country attacked, irrespective of the type of weapon employed; *second*, as long as all the nuclear-weapon powers, that is, powers capable of using nuclear weapons, are also permanent members of the Security Council, any decision concerning military or non-military measures against the delinquent state would require their approval, and it is inconceivable that an aggressor nation would consent to a collective action being taken against itself; and, *third*, immediate active intervention, as envisaged by the resolution, is deemed unacceptable by some nonaligned and neutral states, unless assistance has been specifically requested by the victim. Furthermore, the resolution in question relates to a possible action by the Security Council only when a threat of nuclear attack has been made or the attack has actually occurred. It does not offer assurance for the prevention of the use or threat of use of nuclear weapons. These deficiencies were pointed out by many delegations at the NPT Review Conference. But at the same time, doubts were expressed as to whether it was at all possible in the present world situation to devise such 'positive' security guarantees which would be both credible and effective as well as acceptable to all. There was, therefore, wide support for additional assurances in the form of legally binding 'negative' security guarantees. These were proposed by a group of 11 states: Bolivia, Ecuador, Ghana, Mexico, Nigeria, Peru, Romania, Senegal, Sudan, Yugoslavia and Zaire. The proposal was that a protocol (Additional Protocol III) should be adopted, under which the depositary governments of the NPT would undertake 'never and under no circumstances' to use or threaten to use nuclear weapons against non-nuclear-weapon states party to the treaty whose territories were 'completely free from nuclear weapons', and to refrain from 'first use' of nuclear weapons against 'any other' non-nuclear-weapon state party to the treaty. The protocol would also contain positive assurances patterned after the 1968 UN Security Council resolution, namely, an obligation to provide immediate assistance to a victim of a nuclear threat or attack with nuclear weapons, at the request of the victim, and without prejudice to the obligations under the UN Charter.

Responding to the proposal for 'negative' security guarantees, the USA argued that such commitments, undertaken on a global scale, would not serve the objective of non-proliferation and universal adherence to the NPT; they could encourage those states which are now protected by nuclear-weapon-powers against a threat of a conventional attack, to acquire their own nuclear weapons for defence. In the view of the USA, renunciation of the option of first-use of nuclear weapons would amount to accepting a 'self denying ordinance that weakens deterrence'. It is, therefore, prepared to make use of nuclear weapons 'should we be faced with serious aggression likely to result in defeat in any area of very great importance to the United States in terms of foreign policy'. In this context, reference has been made to possible conflict situations in Europe and in the Korean peninsula, where US tactical nuclear weapons are stationed.

Five states—Ghana, Nepal, Nigeria, Romania and Yugoslavia—submitted a draft resolution which invited the nuclear-weapon states party to the NPT to initiate negotiations on the conclusion of a treaty on the withdrawal from the territories of the non-nuclear-weapon states party to the NPT of all nuclear-weapon delivery systems, especially tactical nuclear weapons.

The sponsors of the protocol on security assurances (Additional Protocol III) requested that the depositary governments should undertake to encourage negotiations to establish nuclear-weapon-free zones and to respect the status of the zones established. Also, in a separate proposal, Iran urged the nuclear-weapon states to undertake a solemn obligation 'never to use or threaten to use nuclear weapons against countries which have become Parties to and are fully bound by the provisions of such regional arrangements'.

The Final Declaration recognized that nuclear-weapon-free zones could contribute to the security of states, but did not specify that the nuclear-weapon states should undertake not to use nuclear weapons against the denuclearized countries.

The Soviet Union and its allies also suggested that the UN Security Council give the force of law, with an internationally binding effect, to the 1972 UN General Assembly resolution on the renunciation of the use of force in international relations and simultaneous prohibition of the use of nuclear weapons. But since the resolution presumed an indissoluble link between non-use of force and non-use of nuclear weapons, it actually condoned the first use of these weapons against any nation in response to a non-nuclear, conventional attack.

Unlike the case of the USA, which long ago officially stated that it was prepared to take 'whatever action with whatever weapons are appropriate' in the event of an aggression that could not be repulsed by conventional forces, the Soviet refusal to accept a no-first-use doctrine is a relatively new development.

The proposals submitted to the Conference reflected the different security situations in which states find themselves. There were clear differences in the attitudes of nuclear-weapon powers, the allies of these powers which believe that they are protected by a 'nuclear umbrella', nonaligned countries which fear a nuclear threat, and nonaligned countries which do not perceive themselves to be under such a threat.

Eventually, the Conference confined itself to issuing an appeal to all states to refrain from the threat or use of force in their mutual relations. This amounted to a reiteration of the UN Charter requirement, valid for all, irrespective of the NPT. But the NPT parties received no assurances from the depositary governments that the weapons they had renounced would not be used against them. It is noteworthy that China, a non-party to the NPT, is the only nuclear-weapon power to declare that it would never, under any circumstances, be the first to use nuclear weapons against any country.

Adherence to the NPT

The Conference expressed the hope that all states that had not joined the NPT would join it at the earliest possible date, but there is no sign that they will soon do so.

By December 31 1975, fourteen countries had signed the treaty but had not ratified it: Barbados, Colombia, Democratic Yemen, Egypt, Indonesia, Japan, Kuwait, Panama, Singapore, Sri Lanka, Switzerland, Trinidad and

Tobago, Turkey and Yemen. Egypt said that it would ratify the NPT only if Israel did the same, and the Egyptian President declared that if Israel obtained nuclear strike capacity, then Egypt would follow suit. In Japan, the government failed to get the treaty ratified by the Diet in 1975. The Indonesian minister for scientific research stated that his government had not ruled out the possibility of developing nuclear weapons. The Turkish defence minister claimed that his country had to have nuclear energy and atomic weapons to 'protect our independence in the present world strategy and to survive', and the Prime Minister made it clear that Turkey would not ratify the NPT.

Those who have neither signed nor ratified the NPT include two nuclear-weapon powers, China and France; India, which has exploded a nuclear device and is engaged in a space programme which may lead to the development of a missile-based nuclear delivery system; and half a dozen states, generally considered as near-nuclear: Argentina, Brazil, Israel, Pakistan, South Africa and Spain. Since Argentina and Brazil have opted to develop nuclear explosives, their adherence to the treaty is improbable.

Even some parties to the NPT have indicated that under certain circumstances they might shed the obligations contracted under the treaty. Thus the Shah of Iran said that if other countries in the region came into possession of nuclear weapons, Iran would also have to acquire them. The President of the Republic of Korea stated that his country would develop its own nuclear weapons, if the US 'nuclear umbrella' were withdrawn. And the Libyan President expressed the opinion that in the future 'atomic weapons will be like traditional ones, possessed by every state according to its potential. We will have our share of this new weapon'. Other countries may have similar ambitions, even though they have not voiced them publicly.

Conclusions

The Conference failed in solving the problems essential for the survival of the NPT. Hence the vagueness and ambiguity of the declaration it issued. The declaration reaffirmed the provisions of the NPT, but ignored the fact that important stipulations were being circumvented. It promised more favourable treatment of the parties, but contained no firm undertakings to end discriminatory supplier policies. It stressed that the responsibilities and obligations of all parties must be balanced, but did not commit the nuclear powers to fulfilling their part of the bargain by reversing the nuclear arms race. The only novel features were the promotion of international arrangements to ensure the physical protection of nuclear materials, and a stimulus to the idea of setting up multinational nuclear fuel cycle centres. Given the rigid attitude of the nuclear-weapon powers, it was perhaps unrealistic to expect more from the Conference. Most other participants expressed deep disillusionment. Yugoslavia even went so far as to announce that it would 're-examine its attitude towards the Treaty and draw corresponding conclusions'.

The next Review Conference is scheduled to take place in 1980. If in the intervening years no progress is made in streamlining nuclear supply policies in accordance with the spirit of the treaty, and if no halt is put to the nuclear arms race, a second meeting of the parties will be faced with a further erosion of the NPT.

5. *The Threshold Test-Ban Treaty**

As a result of the summit meeting of US and Soviet leaders, which lasted from June 27 to July 3 1974, a Threshold Test-Ban Treaty (TTBT) was signed imposing limitations on the two powers' underground nuclear weapon tests. Each party has undertaken to prohibit, to prevent, and not to carry out any underground nuclear weapon test having a yield which exceeds 150 kilotons at any place under its jurisdiction or control, beginning March 31 1976.

What 'sacrifice', if any, have the two powers made by limiting their weapon explosions to a 150-kiloton yield?

Computations made by Swedish seismologists for the period covering 1969–73 show that, for the period examined, most of the US and Soviet explosions were well below 150 kilotons and only a small fraction was above 150 kilotons. Indeed, in recent years, attention has been devoted mainly to warheads for small tactical weapons or for such strategic weapons as have a yield lower than that established in the new treaty. Thus, the restriction imposed by the TTBT does not imply any real sacrifice on the part of the two powers, and the permitted yield is still ten times higher than that of the Hiroshima bomb.

Verification

Article II stipulates that each party will use 'national technical means' of verification at its disposal to provide assurance of compliance with the provisions of the treaty. The same formula was used in the US-Soviet treaty on the limitation of anti-ballistic missile systems (ABM Treaty).

National technical means to control a test ban may consist of seismic monitoring, satellite observation, or electronic eavesdropping. Seismic monitoring is considered to be the most useful method (especially when the size of the tests is to be measured). In a protocol to the treaty, the USA and the USSR have agreed to exchange information necessary to establish a correlation between given yields of explosions at the specified sites and the seismic signals produced, and to improve each side's assessments of the yields of explosions based on the measurements derived from its own seismic instruments.

---

* From *SIPRI Yearbook*, 1975.

Each party undertakes not to interfere with the national technical means of verification of the other party. This clause can be interpreted as an obligation not to use evasion techniques, such as 'cavity de-coupling' involving the emplacement of a nuclear device in a large cavity in hard rock or salt medium which may reduce the recorded seismic magnitude. In addition, the USA and the USSR have pledged themselves to conduct all nuclear weapon tests solely within specified testing areas.

Peaceful nuclear explosions

The provisions of the treaty do not extend to underground nuclear explosions carried out by the parties for peaceful purposes. This means that the threshold of 150 kilotons does not apply to such explosions. But it is not possible to distinguish with certainty nuclear tests serving peaceful aims from those serving military aims.

The parties undertook to conclude 'at the earliest possible time' a separate agreement by which underground nuclear explosions for peaceful purposes shall be governed. An understanding in principle on some of the requirements for verifying that peaceful nuclear explosions are not weapon tests had been reached already in the course of the negotiations on the TTBT. These requirements include prior notification, precise definition of time and place and, most important, the presence of observers.

Summary and conclusions

Unlike the PTBT, the July 1974 treaty limiting underground nuclear weapon tests is a purely bilateral affair between the USA and the USSR. Adherence by other states is not envisaged. Only the United Kingdom announced that it had committed itself to abide by the provisions of the treaty, even though it was not formally party to it. The remaining two nuclear-weapon-powers—China and France—will certainly ignore it, since they have not even subscribed to the ban on atmospheric testing. The majority of other countries have taken a very sceptical view, while India found in the treaty added justification for pursuing its programme of peaceful nuclear explosions.

The treaty has, no doubt, a few positive aspects. It will complicate the development of new high-yield warheads by both sides. It will also make it difficult for them to carry out stockpile-sampling, because the existing large thermonuclear weapons could not be tested at their full yield. Cessation of explosions in the megaton range may have favourable environmental effects. It will reduce risks of radioactive venting, artificial earthquakes or tidal waves. Most important, however, are the verification provisions. It will be the first time that detailed information concerning sites and yield of nuclear weapon explosions will be exchanged between the parties. Apart from the political significance of this procedure constituting a step towards greater openness among states, a suitable verification framework could be created for use in a possible comprehensive test ban. There may also be a 'peaceful' spin-off from the exchange of seimic data. It is interesting to note that Canada has expressed the hope that the seismological and geographical information exchanged between the USA and the USSR, would be made available to all countries, leading, among other things, to a better worldwide understanding of the Earth's structure. Also Sweden asked for access to such data.

Even more significant would be the acceptance of on-site international observation of peaceful nuclear explosions. This would amount to a breakthrough in the great powers' approach, notably that of the Soviet Union, to the problem of verification, and perhaps set a precedent for other arms control measures. Participation of third countries or international organizations in the envisaged observation would further enhance the value of this step.

All this does not alter the fact that the treaty has contributed very little, if at all, to the cessation of the nuclear arms race. No wonder that it was met with disappointment by world public opinion which has condemned all nuclear weapon tests through UN resolutions, and which year after year has been asking for their complete cessation. It can be argued that partial limitation is better than no limitation at all. This would be true if limitation meant some slowing down of the pace of development or deployment of arms. However, in the July 1974 treaty the threshold was set so high that the arms control effect has been entirely lost. The yield limitation does not even reflect the capabilities of the verification methods. According to some scientists, a nuclear explosion of one-tenth the yield agreed upon by the USA and the USSR can be detected by existing means. Other experts have estimated that the detection/identification threshold for underground nuclear explosions in hard rock in the Northern hemisphere lies now in the yield-range of two to three kilotons. The parties are permitted to do whatever they need for the continuation of their nuclear weapon programmes. All warheads for tactical nuclear weapons and most warheads for strategic nuclear weapons, presently deployed or planned to be deployed, can be tested in their full explosive yield.

What can be affected is a possible new generation of nuclear weapons, as it may be difficult to render their yields larger than 150 kilotons without testing. But the value of this constraint is problematic. If absolutely necessary, low-yield tests could be related to devices of larger yields. Besides, the existing trend is to improve the effectiveness of nuclear weapon systems by increasing the accuracy of missiles rather than by increasing the yield of warheads. With the introduction of multiple independently-targetable re-entry vehicles (MIRVs), multi-megaton weapons are falling into obsolescence anyway.

Furthermore, if peaceful nuclear explosions exceeding 150 kilotons were allowed without limitation on their numbers, and without very strict control over their 'peaceful' nature, the US–Soviet treaty may not signify any restriction on testing activities whatsoever.

6. *The Vladivostok SALT II accord**

In a joint statement issued on November 24 1974, the USA and the USSR expressed their intention to conclude an agreement on the limitation of strategic offensive arms, which would cover the period from October 1977 to December 31 1985.

The new agreement will incorporate some relevant provisions of the 1972 Interim Agreement (remaining in force until October 1977) and will include the following limitations: (*a*) both sides will be entitled to have a certain agreed aggregate number of strategic delivery vehicles; and (*b*) both sides will be entitled to have a certain agreed aggregate number of intercontinental ballistic missiles (ICBMs) and submarine-launched ballistic missiles (SLBMs) equipped with multiple independently-targetable warheads (MIRVs).

On December 2 1974, President Ford revealed the agreed aggregate numbers which supplement the general framework of the new agreement. These are: (*a*) for each side a ceiling of 2 400 will be put on the total number of intercontinental ballistic missiles, submarine-launched missiles and heavy bombers; and (*b*) of each side's total of 2 400, the number of missiles that can be armed with MIRVs will be limited to 1 320.

The new agreement would differ substantially from the Interim Agreement, currently in force.

The latter deals only with two types of strategic offensive weapons—fixed land-based intercontinental ballistic missile (ICBM) launchers and ballistic missile launchers on modern submarines. According to an official US statement, the new agreement would include land-based intercontinental ballistic missiles, submarine-launched missiles and heavy bombers, as well as 'certain other categories of weapons that would have the characteristics of strategic weapons'.

Under the Interim Agreement, the freedom to choose the mix of ICBM and SLBM launchers within the agreed overall levels is restricted. The USA and the USSR are not allowed to start (after July 1 1972) construction of additional fixed land-based ICBM launchers, nor to exceed the ceilings either for SLBM launchers—710 for the USA and 950 for the USSR—or for modern ballistic missile submarines—44 for the USA and 62 for the USSR.

Under the new agreement, each power will apparently have greater freedom to determine the composition of its force so long as the aggregate number of strategic weapons does not exceed the set ceiling. The subceiling imposed by the Interim Agreement on land-based launchers for modern 'heavy' missiles will remain in force, while another important subceiling, that on missiles equipped with multiple independently-targetable vehicles, will be introduced.

---

* From *SIPRI Yearbook*, 1975.

The general terms of the Vladivostok statement have yet to be translated into concrete treaty language. Technical details, including definitions, remain to be settled. Measures of verification must be worked out.

Quantitative increases

An important criterion for judging the value of an arms control treaty is whether the fire-power of the contracting parties has been reduced or, at least, frozen at the existing levels. This applies fully to SALT. Thus, the USSR—which in mid-1974 had 1 567 ICBMs, 636 SLBMs and 140 heavy bombers, together 2 343 strategic delivery vehicles—will be allowed to increase the number of its missiles or bombers by 57 to reach the permitted ceiling of 2 400. On the other hand, the USA—which in mid-1974 had 1 054 ICBMs, 656 SLBMs and 420 heavy bombers, together 2 130 strategic delivery vehicles—will be allowed to increase the number of its missiles by 270 to reach the permitted ceiling of 2 400.

The future agreement is to provide for a ceiling of 1 320 MIRVed missiles. The exact number of warheads which will be mounted on each missile is unknown. At present, the US land-based Minuteman III missile contains three warheads, and the submarine-launched Poseidon missile around ten. Soviet MIRVing capability is estimated at four to eight warheads on a land-based ICBM.

On this basis, it would seem safe to assume that a MIRVed missile on each side will have, on average, six warheads. The total number of warheads on the maximum allowed number of MIRVed missiles for each side would then amount to 7 920 (1 320 × 6). In addition, the parties will be allowed to possess 1 080 (2 400 – 1 320) non-MIRVed delivery vehicles. If these carried one warhead each, the aggregate number of warheads for the USA and the USSR would be 9 000 (7 920 + 1 080).

The figures would be considerably higher, if missiles carrying multiple, but not independently-targetable, re-entry vehicles (MRV) were included in the 1 080 limit for non-MRVed vehicles and, especially, if the full payload carried by heavy bombers were to be counted. It would appear that in 11 years, from 1974 to 1985, the USA could increase the number of its strategic warheads by nearly 12 000 and the USSR by over 8 700, and in 13 years, from 1972, when the first SALT agreements were signed, to 1985, by almost 13 500 and over 9 000, respectively. The figures are approximate but certainly not overestimated. They may, perhaps, never be reached. But the important thing is that the parties will not be prevented from doubling or even trebling their present strategic warhead inventories (with yields ranging from about a hundred kilotons to a few megatons), if they decide to do so, and it may be noted that tactical warheads which have a lower yield would not be subject to any restriction whatsoever.

Qualitative improvements

However, these increases are not sufficient indicators of changes in the nuclear fire-power of the parties. At the present stage of the strategic arms race, due to the introduction of MIRVs, it is the number of nuclear warheads, as well as the quality of missiles and warheads, which matter, rather than the number of missiles alone.

Qualitative changes in strategic offensive weaponry over a ten-year period are more difficult to measure. But since modernization and replacement of these weapons will not be prohibited, and may even be expressly allowed as under the Interim Agreement, their accuracy, penetrability, survivability, range and yield-to-weight ratio will, no doubt, improve. In the important field of qualitative improvements in missiles and bombers, including missile throw-weight capability, rivalry between the two powers will not cease. Since the USSR has built heavier land-based missiles than the USA, which permits it to mount a greater number of warheads per missile, the USA may consider increasing the number and/or the yield of warheads on its own MIRVed missiles, if it decides that this is strategically necessary; it would not be prevented from doing so under the new agreement. Neither would the USSR be inhibited in its attempts to match the accuracy of US missiles and in placing greater emphasis on bombers than it has hitherto. The licensing of mobile ICBMs will open new vistas for great-power competitions and may create new dangers to the stability of the strategic balance.

The parties themselves concede that it is not strict strategic equality that they seek, but rather a sort of balance which would give them a perception of equal security. If this is so, the contention that negotiations about reductions would be possible only when the very high levels permitted by the new agreement were reached, seems untenable. It is incomprehensible why a balance and equal security could not be achieved by bringing the present levels, which are already high enough to destroy humanity several times over, down to a common, lower plateau, and by halting or significantly slowing down the introduction of new arms. The policy of continued armaments contradicts the declared intention to disarm. At present, neither side can really threaten the overall strategic forces of the other, while an arms race, by its very nature, generates temptations among the competitors to overtake each other. It is bound to create instabilities and, thereby, new difficulties on the way to weapon reductions.

Consequently, the Vladivostok statement simply rationalizes further arms expansion.

*7. The charade of piecemeal arms limitation**

I am concerned here with why we have so far not even been able to provide the *necessary* progress in Soviet–American nuclear arms control.

* Shortened version of a paper by Bernard T. Feld from *The Bulletin of the Atomic Scientists*, Vol. XXXI, No. 1, January 1975.

Why, after almost 30 years of intensive efforts in this direction, are we nowhere? Indeed, we are even behind where we started, with numbers, types, forms, strengths, and all other destructive aspects of nuclear arms expanding almost unhindered, while the clear understanding that seemed universally to have grown out of the horrors of Hiroshima and Nagasaki—that nuclear weapons must never again be used—seems in grave danger of evaporating from human consciousness.

The international agreements on arms control arrived at since the end of World War II are a far cry from the measures needed to eliminate the nuclear threat.

The rule now seems to be that only those weapons or activities can be eliminated or banned that are of *no* interest to any substantial fraction of the military in the United States or in the Soviet Union.

The Biological Weapons Convention, which goes well beyond the simple no-first-use provisions of the Geneva Protocol of 1925, seems to be an exception to this rule. However, we should note that no serious deployment or attempted use of biological weapons is known to have yet taken place. This probably reflects a corollary to our previously-enunciated rule, namely: No weapon is ever banned (or eliminated by treaty) that has already been seriously deployed by either side.

Now, I would not claim the force of a law of nature for my rules of permitted arms control arrangements. They are obviously simple, empirical extrapolations from previous experience, a mild attempt at establishing some beginnings of a sort-of-sociology of the military aspects of international intercourse. But they serve to describe the limited degree of success of the measures heretofore achieved in curbing the nuclear arms race, and they also help to account for the many failures in earlier attempts as well as for the counterproductive nature of some of the so-called successes.

The first apparent success in the Soviet–American negotiations to control nuclear weapons was the limited or partial test-ban treaty of 1963. But despite the firm pledge of the superpowers to pursue seriously the negotiations toward a universal test ban, the number and magnitude of underground nuclear explosions has grown steadily in the 11 years since the conclusion of the partial test ban. So the pace of weapons development, instead of being inhibited by the treaty, was if anything increased.

Thus, ecological blessing though it is, the Partial Test Ban Treaty has been turned into an arms control disaster. Moreover, the same pattern has been repeated in every important nuclear arms control agreement since 1963.

The results to date of the Strategic Arms Limitation Talks—products of five years of determined bilateral negotiations, of twice as many years of informal and formal technical preparations, and of three summit meetings—add up to a large step backwards.

The ceilings on the total numbers of missile launchers permitted to both sides by the interim SALT I agreement of 1972 were well in excess of the numbers deployed at the time of the agreement; and even within the absurdly high ceilings established in SALT I, the agreement to permit replacement of older missiles by more lethal, newer models has already resulted in a more-than tenfold increase in the over-kill capabilities of both sides.

But, worst of all, the failure of SALT I to come to grips with the problem of curbing new technological advances in nuclear weaponry has been taken as a signal by the military weaponeers on both sides to go full-steam ahead on every new weapons scheme that is not explicitly prohibited by the agreement, no matter how absurdly destabilizing its effects.

So, despite a few positive features—namely the ABM curbs, the mutual acceptance of national verification by means of orbiting satellites and the formalization of the continuing consultation—the main result of SALT has been to initiate a new phase of the nuclear arms competition. Now both sides are moving steadily from their earlier posture of reliance on nuclear weapons for the sole purpose of deterring a first-strike by the other side toward postures that can only be interpreted as preparations for use of these weapons in any conflict against any kind of target, regardless of the actions of the other side. Such an erosion of the inhibitions against any possible first-use of nuclear weapons—so painfully constructed over 26 post-World War II years—unless reversed can only lead to unmitigated disaster.

Three basic flaws

Having painted such a bleak picture, I am obligated to try to understand the root causes of our current dilemma, and to suggest some possible new directions that might reverse the present trends.

In my judgment, there are at least three basic flaws in the piecemeal arms control approach that has thus far been adopted in our attempts to cope with the nuclear danger. Although these flaws may not be fatal, they must nonetheless be overcome if the arms control approach is ever to fulfill its earlier promise of eliminating the dangers of a nuclear holocaust.

The *first* flaw relates to the use of the so-called bargaining chip approach in the negotiating process. Apparently, it has so far proved to be impossible to negotiate away any weapons system that is not yet threatening the other side. As a consequence, both sides believed it to be necessary to accelerate the deployment of weapons systems that they are simultaneously trying to eliminate in the negotiations. But the catch has been that, once deployed, the pressures to retain such systems have proved to be insuperable, even in the face of a successful negotiation, as witness the case of the ABM. So, the best that has been accomplished by each new agreement has been to freeze temporarily the nuclear confrontation at a new level, in each case higher than before the start of the negotiation. This is obviously a losing game.

The *second* flaw relates to the process of ratification of agreements. In order to mollify the opponents of any specific arms control measure, it has until now been necessary to agree to the development and deployment of systems that, in the view of these opponents, tend to redress the balance that has allegedly been disturbed by the agreed-upon limitations. By acceding to such demands, the effect of every agreement to date has been to cause the arms race to burst out in some other direction, frequently a more virulent one than that curbed by the agreement. Because the effect of this flaw has frequently been to permit the development of systems which had previously been rejected as either too crazy or too dangerous, it has been referred to as the 'free-ride' syndrome.

The *third* flaw concerns the time required for the negotiating process, as compared to the time associated with new technological developments in nuclear weaponry. At least until now, the negotiating time has always been greater than that needed to produce new systems, with the result that by the time a weapons system is eliminated via the negotiating process it has already become obsolete. This has had the effect of rendering essentially meaningless almost all past arms control agreements.

Now, it can be logically argued that none of the flaws just enumerated is intrinsic to the nature of the arms control process. The decisions as to whether or not a given system should be developed or deployed, and as to whether restraint or haste should be applied to a given development or to a given negotiation, are political decisions to be made in the political arenas of the countries involved, involving as they do a competition for relatively scarce material and manpower resources. This may be true in theory. But the simple and irrefutable fact is that, given the climates that have prevailed in the competing nuclear powers since World War II, the scales have always been weighted heavily in favor of the advocates of unrestrained technological competition, as opposed to the advocates of technological restraint.

And this has been the case in practically all aspects of technology, which goes a long way toward accounting for the current disarray in the so-called developed world.

Unfortunately, the first law of socio-dynamics seems to be: Whatever is technically possible must be developed. And the second law: Whatever is developed must be used. It may well be that the survival of the human race depends on our ability to repeal these laws.

In any case, whether we are dealing with a basic social pattern or with a set of historical accidents, the sorry history of attempts to control the nuclear arms race is replete with examples of the nullification of determined and honest efforts through the intervention of one or more of these intrinsic flaws. Indeed, the three aforementioned weaknesses in the arms control approach seem to reinforce each other.

A comprehensive approach

Such is the sorry state in which we arms controllers now find ourselves. Having devoted half a life-time to the pursuit of what now appears to have been a will-o'-the wisp, I am forced, reluctantly but inexorably, to the conclusion that the arms control approach simply does not work in today's world—a different approach is essential if nuclear disaster is to be averted. The alternative is, I believe, obvious. Nor does it involve any brilliant, new ideas.

If partial measures have only served to push the competition into new and more virulent directions, the only way out may be through a comprehensive approach. If it has been futile to engage the military technicians in arguments over technical details—to fight gimmicks with gimmicks—perhaps we could have more success by bringing the argument back to matters of principle, to questions of economics, of ethics, of morality and, quite simply, of survival.

Such a program must be based on the realities of the present situation. And what are these realities?

First and foremost is the stark and undeniable fact that the nuclear powers, and especially the so-called superpowers, possess nuclear weapons, and the means of their delivery, sufficient to annihilate each other and much of the rest of the world with them many times over. Irrespective of any measures or countermeasures to which they might resort, a full-scale nuclear war would set the superpowers and the rest of the world back by centuries, if not millennia. Hence, the first essential step is to cut back drastically, to eliminate this over-kill capacity.

Arms control priorities

The first priority of SALT should therefore be to reduce the total number of nuclear weapons and their carriers, not by mere percentages but by factors of two, three, ten.

Once the numbers have been reduced, it will of course be essential to institute controls over new technological developments. But the incentives for controlling new technology under such conditions will be much greater, for control over technology will then be highly relevant to the maintenance of stability. Under today's conditions, it is hard to care less whether the overkill is by a factor of 100 or 1 000.

The second reality is one which was well understood a decade ago, enunciated clearly by Khrushchev and Kennedy, but lost sight of in the pseudo-sophistication of present-day nuclear doctrine. Simply stated, it is that nuclear weapons are useful for one purpose only, and that is to insure that they will never again be used. A return to universal recognition of this reality is essential, not only to prevent nuclear war between the nuclear powers but to provide a necessary basis for the universal adoption of a non-proliferation policy by the rest of the nations.

Toward this end, the nuclear nations must first agree to a policy of no-first-use of nuclear weapons. This doctrine should be clearly and unequivocally enunciated in a no-first-use pledge to which all the nuclear powers must be signatories. Since the People's Republic of China has already on a number of occasions proclaimed its willingness to join in such an agreement, and the Soviet Union as well, the main stumbling bloc, at present is my own country, the United States. It must therefore become a first priority of our arms control community to reverse the present US position on this issue.

The importance of the principle of no-first-use goes beyond the problem of nuclear arms; it goes to the heart of the problem of arms control. If experience has anything to tell us, it is of the importance of drawing a firm and unequivocal line between the use and non-use of any banned form of warfare. The only arms control measure adopted between the two world wars that has survived is the Geneva Protocol prohibiting the first-use of chemical and biological weapons. Any equivocation, any attempt at blurring the line between use and non-use of a controversial weapon, has led to disaster, as witness the horrors of strategic bombing in World War II and more recently the unconscionable consequences of the introduction of supposedly non-lethal chemical weapons by the United States and Indochina.

A third reality in the present situation—and one that poses greater threats than does the possibility of nuclear war between the great powers—is the inevitability of nuclear weapons proliferation, stemming from the widespread proliferation of peacetime fission technology and, particularly, of the fissionable materials used in nuclear energy production. To permit the present laissez-faire approach to continue is to guarantee disaster. Here again, despite all good intentions, the partial, equivocal approach of the present Non-Proliferation Treaty is doomed to failure. With tons of plutonium circulating randomly around the world, there are simply no safeguarding measures that can possibly prevent the diversion of the relatively small quantities required to make an atomic bomb. And, unhappily, the wishful thinking of many so-called experts notwithstanding, it is just too easy for even a relatively small group of moderately competent technicians to fashion a crude bomb out of readily available materials.

To cope with this problem, I believe that we must go back to the original Oppenheimer–Lilenthal approach, or at least to revive some of its most important elements. I am not advocating complete international control over all nuclear activities; this is simply not realistic today. But I do believe that at least as regards the processing and circulation of plutonium, international control is the only acceptable solution. It is certainly not beyond the ingenuity of man to devise a system within which all plutonium (and possibly also highly-enriched uranium) would be separated, processed, shipped, and recovered by a single international agency while, at the same time, guaranteeing equitable access to the supply for all responsible users.

I do not delude myself that the approach I now advocate—or the measures I have suggested—is likely to be universally implemented in the near future. Congenital optimist though I am, I still feel it is more likely than not that nuclear weapons will again be used in my lifetime. I hope and pray that, even if this should occur, the peoples of the world and their leaders will retain enough sanity to prevent such use from escalating into a full-scale disaster, from which there could be no return—that they will recoil from the brink and then take the necessary actions to banish forever the nuclear menace.

But I am even more convinced that there is no hope of escaping disaster short of the recognition by all men of good will that we are facing the gravest threat to mankind's existence, and that nothing short of concerted action can save us.

8. *Nuclear arms control agreements: process and impact**

Any inquiry into the utility of arms control efforts ought to include not only a consideration of the changes in weapons programs that can be immediately attributed to the efforts but also a critical analysis of the impact of those efforts on the weapons acquisition process; on perceptions and doctrine about the role and utility of military force; and on the international climate, broadly defined.

US–USSR nuclear arms control agreements to date have prohibited only a very specifically and narrowly defined range of activities. The favorable reaction to these agreements has stemmed from the sense that they set some precedents and aroused hopes for further more substantial measures to control the strategic arms competition, and some were viewed as major political achievements in this regard.

If the extent of limiting weapons programs is the measure of arms control efforts, the immediate benefits of the post-war years seem limited but not negligible. Some programs have been terminated that otherwise would have almost certainly continued: atmospheric nuclear testing and R&D and stockpiling of biological weapons. Expansion of the arms race into a number of dimensions that are currently of little interest—space, the seabed, and the Antarctic—has been foreclosed, or at least made much more unlikely, as has all work on ABM systems other than fixed land-based ones. As a product of SALT, limits on deployment of ABM systems and intercontinental missile delivery systems will reduce concerns on the part of each of the superpowers about expansion of its adversary's capabilities during the next few years.

The fact remains that there are as yet no examples of the negotiation of measures that clearly affect any of the basic elements of the military forces of the participants. Unlike the interwar naval negotiations, post-war arms control efforts have not sought to limit 'virile' military programs in ways that might substantially change strategic postures. When limits on significant weapons programs have been agreed to, as in the case of ABM and nuclear testing, it has often required placating proponents of weapons programs by leaving them an escape hatch—a narrowed set of dimensions within which to work. R&D, particularly in those areas where production, deployment or testing are prohibited, is seen by many as representing a vital safeguard against unexpected treaty termination and against technical surprises that

* Excerpts from a study by G. W. Rathjens, A. Chayes and J. P. Ruina published by the Carnegie Foundation for International Peace, Washington 1974.

may alter perspectives about an agreement. It is rare to agree to cut off an on going program completely as was the case with biological warfare.

Ironically, efforts to negotiate arms limits involve interrelated processes which may cause retention of arms otherwise expendable or the initiation, acceleration or expansion of weapons programs. We identify three mechanisms that can be of importance.

Symmetry or 'leveling up'

The ABM treaty, in an effort to achieve not only equity but also symmetry, provides an example of retention and permitted construction of some superfluous strategic weapons.

The propensity for near-perfect symmetry is also reflected in attitudes about offensive strategic forces. Demands have already been made that the US must be permitted to match the currently superior total throw weight of the Soviet missile force. Likewise, it is to be expected that the Soviet Union will insist on being permitted to match US multiple warhead capability and perhaps also the US long-range bomber force size. Achieving symmetry will require either asymmetrical reductions or permitted increases on both sides to the highest common denominator. And though in the absence of negotiations each might have little interest in emulating the other, the commitment engendered in the negotiating process may make it likely that any agreed ceiling will become a floor. If the past pattern of behavior is predictive, the Soviets will be permitted a MIRV capability comparable to that of the US rather than the US rolling back its MIRV deployments, all by agreement.

The impact of intragovernmental negotiations

This leads us to a discussion of a more troublesome driving force in the arms race—the negotiations within each government that are associated with arms control efforts. There is a natural propensity of political leaders to buy off potential critics by agreeing to demands made by them or at least by meeting them half way.

One prominent example of internal bargaining in the United States is to be found in the test ban safeguards. When the Treaty was under consideration, the President, in response to opposition pressure sent a letter to the Senate leadership promising:

(*a*) The conduct of comprehensive, aggressive, and continuing underground nuclear test programs designed to add to our knowledge and improve our weapons in all areas of significance to our military posture for the future.

(*b*) The maintenance of modern nuclear laboratory facilities and programs in theoretical and exploratory nuclear technology which will attract, retain, and insure the continued application of our human scientific resources to these programs on which continued progress in nuclear technology depends.

(*c*) The maintenance of the facilities and resources necessary to institute promptly nuclear tests in the atmosphere should they be deemed essential to our national security or should the treaty or any of its terms be abrogated by the Soviet Union.

(*d*) The improvement of our capability, within feasible and practical limits, to monitor the terms of the treaty, to detect violations, and to maintain our knowledge of Sino-Soviet nuclear activity, capabilities and achievements.

In consequence there was a weapons test program that involved a substantial increase in the number of nuclear tests as compared with the pre-test ban period, the average number per year since the 1963 test ban being about twice that of the early fifties.

In the case of the SALT agreements, the Joint Chiefs also had their conditions for concurrence, this time three, which they referred to as 'assurances'. They argued that in order 'to guard against a degradation of its national security posture' the United States must have the following:

> *Assurance I.—A broad range of intelligence capabilities and operations to verify Soviet compliance in a strategic arms limitation environment.*
>
> *Assurance II.—Aggressive improvements and modernization programs.*
>
> Maximize strategic capabilities within the constraints established by the ABM treaty and the interim offensive agreement.
>
> Plan for rapid augmentation of strategic forces beyond the constraints of the treaty and agreement to be made in the event of abrogation, withdrawal, or collapse of negotiations.
>
> *Assurance III.—Vigorous research and development programs.*
>
> Maintain testing systems technological superiority.
>
> Continue testing to insure the effectiveness of new and existing nuclear weapons systems.

In a somewhat attenuated sense this is happening. The B-1 bomber program and the Trident missile-launching submarine programs are continuing, the latter somewhat accelerated, and after declining for several years, the R&D budget for strategic arms is increasing. Perhaps the extreme example of responsiveness to the spirit of the second assurance is to be found in the immediate efforts by the Defense Department to initiate a program to develop a new cruise missile to be launched from submarines. The 'safeguards' or 'assurances' are only the most obvious manifestations of the effects of intragovernmental bargaining that are pervasive in any international negotiating situation.

Bargaining chips

This brings us to what may be the most significant escalatory process involved in efforts to limit arms—the need for 'bargaining chips'. It is hardly a new notion. The material relating to the naval limitations efforts of the inter-war years is replete with references to the need for active shipbuilding programs if a nation wished to be in a position to bargain effectively.

New weapons programs conceivably could originate in efforts to acquire 'bargaining chips'. More commonly, old ones that are wanted primarily for quite different reasons will be sold to otherwise dubious decision-makers on the grounds that they are required if future negotiations are to be successful. SALT provides two dramatic examples.

For over a decade, proponents, particularly in the Army, tried to secure an affirmative decision on a US ABM deployment. There can be little doubt that some Senators by 1970 had serious doubts about the technical and military effectiveness of the system, but supported continued deployment primarily because they believed the Administration's argument that it was needed to secure a satisfactory SALT outcome.

A very similar situation has arisen with respect to SALT II. In this case the Trident submarine program has been the 'chip'. While R&D on Trident has commanded wide Congressional support, there has been substantial sentiment that decisions relating to production and deployment are premature and may eventually prove unwise. The major point we would make here is simply that the program got the support it did in the Congress in 1972 and 1973 in part because of acceptance of the Administration's claim that it was needed so that we could negotiate from a strong position at SALT II.

Although attempts have been made by arms control proponents to discredit the 'bargaining chip' approach to negotiations, one can hardly doubt its importance. And to look at the question from the other side, it seems clear that the US desire to negotiate an offensive forces agreement and its willingness to settle for one that has been severely criticized as being unfavorable to the US was strongly influenced by the Soviet Union's impressive 'bargaining chips'—intensive deployment programs for both ICBMs and SLBMs.

From an arms control perspective there are two major problems as regards 'bargaining chips'.

First, an adversary may acquire additional 'chips' of his own to avoid having to negotiate from a position of weakness. Thus, the action-reaction phenomenon, which is held by many to be an important factor in arms races, may take on enhanced significance in a negotiating situation.

Second, a new program once underway, even if justified as a 'bargaining chip', acquires a constituency likely to grow in strength and sense of commitment, that has a stake in its continuation. Consequently, if agreement is ever reached to limit the weapons in question, it is very likely to involve their retention, and, as noted earlier, because of symmetry considerations, possibly the acquisition of similar weapons by an adversary. It is particularly unlikely that 'bargaining chips' will be cashed if the weapons in question have involved great expense, if there is a real belief that they are possibly militarily effective ones, and if the 'bargaining chip' argument was used to tip the balance in favor of a program that already had the backing of powerful groups. The Trident program would fit the description.

The longer the negotiating process, the more troublesome these problems are likely to be.

The 'displacement' effects of arms control efforts

We turn now to a more ambiguous area: displacement effects. The negotiating process and any agreements reached can encourage displacement of resources into new channels which may or may not serve the purpose of arms control. Agreements are reached that foreclose continued efforts in some areas, so resources, both financial and technical, become available for exploiting possibilities not foreclosed and which at the time may not even be foreseen. A number of examples can be cited going at least as far back as the Treaty of Versailles.

That Treaty prohibited construction by Germany of naval warships with displacement in excess of 10 000 tons. A consequence was the design of 'pocket battleships', which were ostensibly within the tonnage limit but which had capabilities, notably in the guns carried, greatly in excess of other ships of that size.

It has recently been suggested that the development of multiple warheads for the Polaris A-3 missile may be an example of the displacement effects. It is argued that the A-3 was to a substantial degree the result of an arms control effort, the informal moratorium on nuclear weapons testing of 1958–61. A single warhead in the megaton range could have been used for the new missile. Two options were available: one, a scaled up but untested version of a smaller weapon; and the other, the same design being used for Minuteman missiles. Neither was judged acceptable, the first because of reluctance to deploy an unproved weapon, and the second because of unwillingness to have too large a fraction of the US missile force equipped with warheads that might have the same vulnerabilities (not to mention possible reluctance of the Navy to use an Air Force warhead). So a decision was made to use three warheads on each A-3 missile, with damage-producing effects about the same as if a single larger warhead were carried. It is not clear that the A-3 development was a strong factor leading to MIRV development, but it may have contributed to the establishment of a climate in which MIRV work commanded support. If so, this obviously was a totally unexpected consequence of the nuclear test moratorium. It is easy to imagine a number of displacement effects of SALT.

In addition to redirecting resources into weapons programs that would not otherwise be funded, there is the possibility that agreement may be a factor in the acceleration of existing programs. If an agreement proscribing some activity seems likely, there will be pressure to make as much progress as is possible while it is still permitted. One is reminded of the flurry of activity that characteristically occurs during ceasefire negotiations as parties seek to establish control over as large an areas as is possible before agreements enter into effect.

Conclusion

The several post-World War II arms control agreements between the United States and the Soviet Union, together with the attendant negotiations, represent a very large investment of intellectual and political energy. Yet the striking fact is how little either the hopes of their supporters or the fears of their opponents have been realized.

The arms control community has been disappointed by the relatively small impact these efforts have had, in sum, on military forces and military budgets. Broadly speaking, there appear to be two reasons for this.

First, individual military programs acquire strong constituencies and great momentum. The forces opposing them are relatively very weak. Thus, arms control 'successes' have been heavily biased toward proscription of work in areas to which there was little commitment at the time of negotiations. This means they are perhaps best viewed from a military perspective as self-denying ordinances. The outer space, Antarctic, seabed, and Nuclear Non-Proliferation treaties are examples. Even the ABM treaty has something of this quality, since technical and economic considerations made large-scale ABM deployment unattractive. The Nuclear Test Ban Treaty may be something of a special case in that the pollution issue was a major factor in the opposition to continued testing.

Second, even where agreements impose severe constraints on specific programs—the Nuclear Test Ban Treaty is illustrative—there is not likely to be a very far-reaching effect on the arms race. What happens is that resources are channeled into somewhat different directions: in the cited case, into underground testing. In fact, because of this phenomenon, and because the pace of negotiations is slow compared with that of weapons acquisition, it seems unlikely that any aggregation of specific constraints of the kinds negotiated so far can by themselves bring the arms race under control. Moreover, given the requirements for wide consensus among sharply divergent interests within the government involved, it seems unlikely that very much broader or more generalized limitations can be agreed to.

Thus, if we view strategic arms control efforts from a strictly military perspective, we must conclude that their function has been a narrow one. The agreements limit concerns, very likely exaggerated, about military developments in the precluded areas; but they do little beyond that. They serve to define and formalize the range within which competition is to be permitted to develop—on the whole in relatively familiar and established directions.

As to the effects of arms control efforts on international relations, the series of efforts we have examined could be seen as mainly registering the profound readjustment in the mutual perceptions and relations of the two superpowers over the last decade. However, we would argue for a somewhat more favorable appraisal. Even the least important of the efforts has served to bring the superpowers together at times when communications were difficult and relations strained. More, the several arms control efforts may actually have operated to some degree as vehicles for bringing about change.

Certainly the most dangerous aspect of the cold war has been the possibility that it might be translated into nuclear terms, and thus it was almost inevitable that any devolution from the phase of most intense hostility should come in part through a consideration of nuclear weapons issues.

It is possible that in the decade ahead political and military leaders will realize that nuclear war and preparations for it are largely irrelevant to the settlement of international differences. If this occurs, it will be possible to approach nuclear arms control issues with less concern about the balance of strategic forces and the likelihood of their deliberate use, and with more attention to the other adverse effects of maintaining them: the costs, the risks of accident, miscalculation and unwanted escalation, and the enormous damage should they be used. In that case we could begin some actual disarmament.

*9. Mutual force reductions in Europe**

On October 30 1973, after five months of preparatory consultations, the representatives of 19 NATO and WTO states began negotiations in Vienna on the mutual reduction of forces and armaments and associated measures in Central Europe.

The present military balance between NATO and WTO forces is, without doubt, of immense importance to the outcome of the current negotiations. Since the mutual force reduction negotiations are primarily concerned with conventional forces, this brief assessment will focus mainly on these forces. However, the military balance in Europe can be assessed not on the basis of NATO and WTO conventional forces alone, but also on how one views the relationship between strategic nuclear and conventional forces in an era of East–West nuclear parity. It is well known that the USA and the USSR have enough strategic nuclear weapons to inflict untold damage on each other. The threat posed by the enormously destructive nuclear forces of both East and West, along with the risk of escalation of any major conflict to the nuclear level, would appear more than sufficient to deter any aggression involving massive conventional forces. But on the other hand, awareness of the enormous destructive effects of nuclear weapons has led to changes in the military strategies of both alliances allowing for the

* From *SIPRI Yearbook*, 1974.

possibility of waging local wars with conventional forces. As a result, increased emphasis has been put on the role of conventional forces in an era of relative balance in strategic nuclear weapons. Even though there is no agreement about the current state of balance or imbalance, it might perhaps be possible to say that there is a rough parity between NATO and WTO conventional forces, all factors taken into account, although it is impossible to express this total conventional balance in numbers and/or percentages.

There are several difficulties in comparing NATO and WTO forces. Some elements of the forces are quantifiable and constitute more or less adequate indicators of military strength in their respective fields. But there are also many other factors which must be taken into account to make a fair assessment of overall military strength. Theoretically these factors can be divided into the following categories: (*a*) other quantifiable elements which are roughly comparable, but for which available information does not permit exact estimates; (*b*) quantifiable elements which appear comparable and for which estimates are available, but which are not actually comparable when their constituent parts are considered; (*c*) quantifiable elements on each side, for which estimates may or may not be available but which have no counterpart on the opposing side; and (*d*) non-quantifiable factors with regard to which relative advantages can be assessed only very roughly, if at all. It is obvious that these factors cannot simply be added together to form an objective measure in mathematical terms of total military strength. Such factors include, for instance, figures for manpower and equipment of various kinds; considerations of geographical advantages, deployment, training and logistics support; and differences in doctrine, philosophy and the like.

Even if it were possible to give comparative values to all the relevant aspects of the forces to be compared—and it is not possible—the problem of assigning a relative weight to each aspect in the overall assessment would still remain. The importance attached to different elements of the forces must be based mainly on assumptions about developments in the event of a war. Such assumptions can be neither proved or disproved, nor, in most cases, shown to be more or less plausible or likely. Problems of this sort arise even when assessing the peacetime balance of forces, to which most comparisons are limited, but they are more difficult when the question of reinforcements is introduced. Finally, existing estimates of the forces of NATO and the WTO are primarily based on Western sources and, as such, inevitably reflect more or less one-sided assessments.

All these difficulties clearly show that any assessment of the military balance between NATO and WTO involves comparisons of so many non-quantifiable and unpredictable factors that it is virtually impossible to make a meaningful overall evaluation. Consequently an assessment of parity is simply another way of saying that there are advantages and disadvantages on both sides.

Bearing in mind the proposals of NATO and WTO countries put forward before the negotiations started and the views expressed by the participants at the beginning of the negotiations in Vienna, one can conclude that the major issues on which agreement will be sought are: (*a*) methods of reduction; (*b*) types of forces to be reduced; (*c*) the territory to be covered; and (*d*) 'associated measures' which may be adopted.

Methods of reduction

Differences exist between the two alliances on the question of the methods which should be applied to reduce existing forces. Different kinds of reductions have been proposed such as 'balanced', 'symmetrical', 'asymmetrical' and so on.

The concept of so-called balanced reductions has caused difficulties. The United States and other NATO countries advocate that 'mutual reduction should be reciprocal and balanced in scope and timing'. The NATO countries claim that there are significant, 'objective' disparities affecting the current situation in Central Europe. According to Western estimates these disparities are in manpower, in the character of forces and in geography. As far as manpower is concerned, the NATO countries consider that the countries of the WTO have more ground personnel on active duty in Central Europe than does NATO. With respect to the character of the forces, the WTO forces maintain a concentration of heavy armour in Central Europe. A marked imbalance in tanks therefore exists in Central Europe. The West also claims that the WTO has superiority in air defence, in both radar detection and interceptors, and that it has about twice as many fighter aircraft as NATO.

This 'imbalance' is even greater when mobilization and the reinforcement of existing forces are taken into account. It is assumed that the WTO would have an advantage here within the first few weeks and that subsequently NATO would be able to reinforce at an increasing rate.

Better reinforcement facilities are primarily based on the 'geographical factor'. Generally speaking, Western countries consider that the geographic advantage is clearly on the Soviet side for two basic reasons: first, because the proximity of Soviet territory to the Central Region makes it much easier for the Soviet Union, than for the USA, to reinforce its troops there and to maintain lines of communication with the base, and, second, because NATO countries lack sufficiently deep territory in Europe to provide for maximum manoeuvrability of forces and adequate defence.

The perceived geographical advantage on the WTO side has been advanced as a major argument supporting NATO proposals for an asymmetrical approach with a view to preserving the current balance. Of course, asymmetrical reductions, while justified by geographical factors, would also reduce WTO superiority in certain quantitative elements of the forces. This could be seen either as improving the current situation from the NATO point of view, or as creating a new situation which would be more balanced than the present one.

The WTO countries, on the other hand, advocate that the reduction of armed forces and armaments in Europe should be based on the principles of 'parity reduction' and undiminished security for both sides. They reject the principle of 'balanced' reduction since, as it is interpreted by the West, it would lead to asymmetrical reductions which are clearly disadvantageous to the WTO position. It has also been argued by the East that the principle of balanced reduction is based only or primarily on factors and elements which show a superiority on the part of WTO forces, while other elements, showing the opposite, are disregarded.

Concerning the reference to the fact that, because of its geographical proximity, the Soviet Union could more quickly mobilize and transfer its divisions to Central Europe it is pointed out that 'NATO's military command has at its disposal a wide range of transport facilities, an intricate network of airfields, the extensive communication system of Western Europe, all the various NATO transport aircraft and so on'.

Taking into account all these components, the WTO members consider that there is a balance between the armed forces and armaments of the two alliances and that any asymmetrical reduction would create one-sided military advantages and as such cannot represent a sound position for negotiations based on the principle of equal security.

But although the West insists on so-called balanced reduction, the two sides have moved somewhat closer. This is illustrated by the Soviet–US communiqué of May 29 1972 in which it is pointed out that 'any agreement on this question should not diminish the security of any of the sides'. During the preparatory consultations the concept of balanced reduction was questioned by the WTO countries and as a result of their opposition, the word 'balanced' does not appear in the Final Communiqué of June 28 1973 in which the subject of the negotiations is defined as 'mutual reduction of forces and armaments and associated measures in Central Europe'.

At the Vienna negotiations on MFR, although both sides have accepted as a point of departure the principle of undiminished security and are advocating the reduction of forces in such a way as to lead to a more stable military balance at lower levels of forces, they differ in the way they interpret these principles, and as a result the differences between them on the question of the kinds of reduction they envisage have not been resolved.

Foreign and national forces—conventional and nuclear

The next question is which forces should be reduced. Both sides were ready to discuss reductions of foreign as well as indigenous forces, while it remained unclear whether these reductions should be carried out simultaneously or successively.

The Western plan called for cuts of about 15 per cent each by US and Soviet forces stationed in Central Europe, to be followed by the setting of a common ceiling for all NATO and WTO forces in the region. By contrast, in the statements and proposals of the WTO countries, reduction of both national and foreign forces and armaments is envisaged.

Another problem connected with the question of reduction of forces is that concerning what kinds of troops and armaments should be reduced, that is, whether the reduction will be confined only to ground forces and conventional weapons or whether it will also embrace air forces and tactical nuclear weapons.

In the opinion of the Soviet Union and other WTO members, reduction of armaments cannot be contemplated without proper reference to the nuclear weapons at the disposal of the NATO allies and in particular of US forces deployed in Europe. The reduction of conventional forces and weapons only partly contributes to strengthening security in Europe.

On the other side it is the firm position of the United States and its allies that, in view of Soviet geographical advantages which provide for easier redeployment of forces probably within a shorter period of time, continued reliance on nuclear weapons is a reasonable guarantee that reduction of forces would not operate to the military disadvantage of either side.

It is further stressed that the supposed conventional force superiority of WTO could be best matched by an adequate supply and use of tactical nuclear weapons.

With respect to the territories where the reduction of forces and armaments should take place, disagreement between the two alliances has been partly resolved. Agreement was reached between the two sides that 'during negotiations, mutual reduction of forces and armaments and associated measures in Central Europe would be considered'.

Associated measures

As has already been mentioned, in addition to reduction of armed forces and armaments, so-called associated measures are also the subject of the current negotiations in Vienna.

Judging by the previous proposals on disarmament and arms regulations in Europe and by studies published on the matter, as well as by opening statements made by the participants to the negotiations, it can be concluded that this term includes three different kinds of measures: confidence-building measures, reduction or freezing of military budgets and verification procedures.

The representatives of NATO countries talk about measures affecting military activities. Although NATO participants have not specified these measures it is clear that by such measures they mean so-called confidence-building measures. These measures are usually understood to include advance notification of military manoeuvres, exchange of observers by invitation at military manoeuvres, prohibition of manoeuvres in border areas, inspection against surprise attack and similar measures.

Although the participants to the negotiations on MFR have up to now not included the reduction or freezing of military budgets in their proposals, the possibility of bringing this question into focus should not be excluded.

If the question of reducing or freezing military budgets is discussed at the current negotiations, one of the central problems will be determining, for each country, the relationship between internal accounting procedures for the financing of military activities and the definition of military expenditure agreed upon. This will be essential for verification since, for many countries, a reasonably comprehensive definition of military expenditures will include activities not financed under the formal defence budget. A second major issue is likely to be whether military expenditures are to be presented in current or constant price terms. In the latter case the treatment of inflation becomes important. Bearing in mind the very complicated nature of the issues involved in such an undertaking it would be necessary to make a thorough analysis before suggesting possible approaches to the problem.

The third kind of measures which are more of a supplementary character are those related to verification procedures after an agreement on force reduction is reached. Although these questions have been raised only by NATO participants it can be assumed that all participants to the negotiations are interested in adequate assurance that the terms of an agreement on reduction were being faithfully carried out and that each party to a possible agreement would refrain from any action which would circumvent or undermine the agreement.

Agreed principles

On the basis of the agreement reached at the preparatory consultations and the proceedings of the current negotiations some conclusions can be drawn about the attitudes of states participating in the negotiations and about the measure of agreement which has been accomplished.

At the preparatory consultations, a number of principles were agreed upon as guidelines for the negotiations. (1) Participants are divided into two categories: those with decision-making power who are direct participants and whose territories are within the agreed area of Central Europe as well as those direct participants who have their troops within the mentioned area (Belgium, Canada, Czechoslovakia, the Federal Republic of Germany, the German Democratic Republic, Luxembourg, Netherlands, Poland, the USSR, the United Kingdom and the USA; and participants with a special status (Bulgaria, Denmark, Greece, Hungary, Italy, Norway, Romania and Turkey). (2) The subject matter of negotiations is defined as the 'mutual reduction of forces and armaments and associated measures in Central Europe'. This lengthy formula was a compromise between different formulas used by two sides. The essential elements in the agreed formula are that reductions would be mutual, focused on Central Europe and accompanied by 'associated measures'. Since no agenda was accepted, the agreement was reached that 'any topic relevant to this subject matter may be introduced for negotiations by any of those states which will take the necessary decisions'. (3) The general objective of the negotiations is defined as 'a more stable relationship' and 'the strengthening of peace and security in Europe'. (4) 'The principle of undiminished security for each party' is defined as a basic principle of the negotiations. (5) Such procedural rules as those concerning the confidentiality of meeting and documents, the rights of all participants to speak and circulate papers on the subject matter, official languages and so on shall apply to the negotiations themselves.

*10. Confidence-building measures in Europe**

The first part of the Final Act of the Helsinki 1975 Conference on Security and Cooperation in Europe (CSCE) dealing with security questions in Europe, contains a 'document on confidence-building measures and certain aspects of security and disarmament'. This chapter of the Final Act—the only one directly related to military issues—consists of a preamble and the following sections: prior notification of major military manoeuvres; prior notification of other military manoeuvres; exchange of observers; prior notification of major military movements; other confidence-building measures; questions relating to disarmament; and general considerations. The document was the result of the work of a special military committee of the Conference, which devoted as many as 246 formal and innumerable informal meetings to reaching a compromise.

Prior notification of major military manoeuvres

Notification will be given through the 'usual diplomatic channels' to 'all' states participating in the CSCE, and not merely to states having a common frontier with the country responsible for the manoeuvres, as was proposed by some delegations at the early stages of negotiations.

Although notification is to be given only for major military manoeuvres, the term 'major' has not been defined. Nevertheless, it follows from the text that it means a manoeuvre involving at least 25 000 troops (a compromise between 12 000 and some 45 000 troops, originally suggested by the Western countries and the Soviet Union, respectively). These include land forces and, in this context, also amphibious and airborne forces, engaged in independent exercises or in combined exercises with any possible air and naval components. Notification 'can' also be given in the case of combined manoeuvres which do not reach the above total, but which involve land forces together with 'significant numbers' of either amphibious or airborne troops, or both. The idea behind the latter provision is that manoeuvres with fewer than 25 000 troops could be considered as 'major' if they involved troops especially trained for invasion purposes. Manoeuvres of naval and air forces, conducted independently or combined with each other, are not

* From *SIPRI Yearbook*, 1976.

covered, presumably because such forces alone are not suitable for occupying territory.

Manoeuvres taking place within Europe (that is, anywhere in Europe, not just within a specified distance from the border of the neighbouring states, as was first proposed by the Soviet Union) and in the adjoining sea and air space are subject to notification. The territories of the USA, Canada and of the Asian part of the USSR are not covered. While the geographical limits of the European territory did not give rise to particular controversies (the USSR agreed to include Georgia and Armenia, and Turkey agreed to include Anatolia), the term 'adjoining' used in conjunction with sea and air space was discussed at length and finally left in the text without its meaning being clarified. In the understanding of many participants it includes the Mediterranean.

A separate formula has been devised for states whose territories extend beyond Europe (the Soviet Union and Turkey). Prior notification by the state in question needs to be given only of manoeuvres which take place in an area 'within 250 kilometres from its frontier facing or shared with any other European participating State'. A proposal to count this distance from the frontier of the 'other participating state', instead of counting it from the frontier of the state whose territory extends beyond Europe, was found unacceptable, because in the case of sea borders (such as between Sweden and the USSR), a good part of the area covered by notification would include international waters.

Notification is to be given 21 days or more in advance of the start of the manoeuvres (a compromise between five and 60 days, originally requested by the USSR and the UK, respectively) or, in the case of a manoeuvre arranged at shorter notice, at the earliest possible opportunity prior to its starting date.

As regards the content of the notification, the document requires the following information to be provided: the designation, if any, the general purpose of and the states involved in the manoeuvre, the type or types and numerical strength of the forces engaged, and the area and estimated time-frame of its conduct. 'If possible', the states will also give additional information, particularly that related to the components of the forces engaged and the period of involvement of the troops, which is a larger notion than the time-frame of the manoeuvres themselves.

Prior notification of other military manoeuvres

The participants in the conference recognized that notification of 'smaller scale' military manoeuvres could also contribute to strengthening confidence. Indeed, in certain areas of Europe, manoeuvres of troops even considerably below the 25 000 level may cause alarm, in particular when they are conducted in the vicinity of other countries. Special emphasis has, therefore, been placed on the advisability of notifying states which lie near the area of such manoeuvres.

Another paragraph recognizing that states 'may notify other military manoeuvres conducted by them' can be interpreted as indicating the desirability of notifying, for example, independent naval and/or air force manoeuvres.

Exchange of observers

The participating states will, if they choose, invite observers to attend military manoeuvres. This stipulation is not linked with the provision on notification of major military manoeuvres. Thus, also other manoeuvres could be observed. Invitations would be issued on a 'bilateral basis' and 'in a spirit of reciprocity', that is, not necessarily to all states.

While it would be up to the invited state to designate its observers, the number of observers as well as the procedures and conditions of participation would in each case be determined by the inviting state.

Prior notification of major military movements

Under this heading the document declares that states 'may, at their own discretion', give notification of their major military movements. (Again, the term 'major' has not been defined.) It contains a promise that further consideration will be given to the question of such notification, 'bearing in mind, in particular, the experience gained by the implementation of the measures which are set forth in this document'. A scant treatment accorded to major military movements is not surprising: some states contend that their security would be adversely affected, if they undertook to give notification of such activities. But it is certainly regrettable, because transfers of combat-ready army, naval and air force units outside their permanent garrison or base areas, with the purpose of redeployment, may cause greater concern than manoeuvres.

Other confidence-building measures

Another measure on which the participants agreed is the promotion of exchanges 'by invitation' among their military personnel, including visits by military delegations, a practice which has existed for years among nations maintaining normal relations.

The participants have also undertaken to take into account and respect the objective of confidence-building when conducting their military activities in the area covered by the provisions for the prior notification of major military manoeuvres. This is a weak reflection of the proposal made by nonaligned countries that states should refrain from any activities by their armed forces which are liable to cause misunderstandings and tension (a euphemism for provocations with a view to exerting political pressure and intimidating neighbouring states).

Sweden submitted a proposal for greater openness in the presentation by the participants in the CSCE of statistics concerning defence expenditures. The initiative was intended to allay misconceptions about military efforts of states and, thereby, contribute to confidence-building. It was not accepted under the pretence that the Conference had no mandate to consider questions which were of a global rather than regional nature.

Questions relating to disarmament

While recognizing the interest of all the participating states in 'lessening military confrontation and promoting disarmament', the document contains no concrete proposals for arms control in Europe. Neither does it give an indication as to what kind of measures would be necessary in the first place. It uses the same phraseology about the need to achieve general and complete disarmament as is contained in UN resolutions adopted yearly since the late 1950's.

General considerations

In this concluding section, three points are worth mentioning: *first*, the recognition of the relationship between security in Europe and security in the Mediterranean area (this was due to the contributions received, and statements heard at the Conference, from Algeria, Egypt, Israel, Morocco, Syria and Tunisia; *second*, the importance attached to the provision of information about developments, progress and results achieved in negotiating fora; and *third*, the acknowledgement of the justified interest of states not participating in these fora to have their views considered. It remains to be seen whether the latter two points will result in rendering the bloc-to-bloc talks on reductions of forces in Europe less esoteric than heretofore, and in reducing the secretiveness surrounding them.

Conclusions

As can be seen from the above exposition, most provisions of the document on confidence-building in Europe are vague and non-committal. The only provision which is couched in concrete terms is that concerning notification of major military manoeuvres. In assessing it one should bear in mind that this is not a legally binding commitment; as stated in the preamble, the measure envisaged 'rests upon a voluntary basis'. Nevertheless, it is a declaration of intention solemnly adopted by the representatives of the participating states at the highest possible level. The undertaking to notify can, if scrupulously implemented, signify a modest first step towards openness in military affairs and help to disperse the myth that secrecy is necessarily an asset for the security of states.

11. *Nuclear-weapon-free zones**

One of the approaches to the nuclear arms-control problem, which has been much discussed since the mid-1950's, is the so-called zonal approach.

* From *SIPRI Yearbook*, 1976.

The idea was conceived with a view to securing the absence of nuclear weapons in certain regions of the world, outside the territories of the nuclear-weapon powers. Several treaties concluded in recent years already reflect this concept. Thus, the Antarctic Treaty of 1959 established a demilitarized régime in the Antarctic, which prohibits the introduction of nuclear weapons into the area. The 1967 Outer Space Treaty contains an undertaking by the parties not to place in orbit around the earth any objects carrying nuclear weapons, nor to install such weapons on celestial bodies. The 1971 Sea-Bed Treaty forbids the emplacement of nuclear weapons on the sea-bed and the ocean floor and in the subsoil thereof, beyond the outer limit of a 12-mile sea-bed zone. The common feature of these treaties is that they concern uninhabited areas where no country exercises national sovereignty, or where, as in the case of Antarctica, claims to territorial sovereignty have not been generally recognized. The only international instrument which has established a nuclear-weapon-free zone in a populated area is the 1967 Treaty of Tlatelolco for Latin America. But since, at the time of its conclusion, no country in the region possessed nuclear weapons and no nuclear weapons were deployed there, the treaty had merely legalized the existing situation. It falls under the same category of preventive measures as the Antarctic Treaty, the Outer Space Treaty and the Sea-Bed Treaty.

Most denuclearization proposals put forward over the years in various international bodies relate to areas where nuclear weapons are already stationed; they imply, therefore, a withdrawal of these weapons. This is the case of Central Europe, the Balkans, the Mediterranean, the Indian Ocean and the South Pacific. Other proposals involve areas where the presence of nuclear weapons is suspected by neighbouring countries, as in the Middle East (Israel), or where the manufacture of nuclear weapons is considered imminent, as in South Asia (India), or Africa (South Africa), or where the introduction of nuclear weapons in case of war remains a possibility, as in Northern Europe. None of these ideas has become the subject of negotiations. As a rule, the proposals are deemed disadvantageous or unfair to one side or another. In addition, the great powers are reluctant to accept restrictions on the deployment of their nuclear weapons, which might affect their global strategic interests.

Nonetheless, in 1974, on the initiative of Finland, the UN General Assembly decided that a comprehensive study of the question of nuclear-weapon-free zones 'in all of its aspects' should be undertaken. This task was to be carried out by an *ad hoc* group of 'qualified governmental experts' under the auspices of the CCD. The group met from June 23 to August 18 1975.

The report of the governmental experts

The report produced by the group describes the concept of nuclear-weapon-free zones; attempts to define the responsibilities of states within the zones and those of other states; suggests verification and control measures;

discusses the relationship of nuclear-weapon-free zones with international law, existing treaties and the United Nations; and deals with the peaceful uses of nuclear energy in the context of denuclearization. But consensus was reached on only a few rather trivial, self-evident principles. The rest of the report is a compilation of contradictory views on matters most essential for the realization of the nuclear-weapon-free zone concept.

The principles agreed upon are as follows:

> Obligations relating to the establishment of nuclear-weapon-free zones may be assumed not only by groups of States, including entire continents or large geographical regions, but also by smaller groups of States and even individual countries;
>
> Nuclear-weapon-free zone arrangements must ensure that the zone would be, and would remain, effectively free of all nuclear weapons;
>
> The initiative for the creation of a nuclear-weapon-free zone should come from States within the region concerned, and participation must be voluntary;
>
> Whenever a zone is intended to embrace a region the participation of all militarily significant States, and preferably all States, in that region would enhance the effectiveness of the zone;
>
> The zone arrangements must contain an effective system of verification to ensure full compliance with the agreed obligations;
>
> The arrangements should promote the economic, scientific, and technological development of the members of the zone through international co-operation on all peaceful uses of nuclear energy;
>
> The treaty establishing the zone should be of unlimited duration.

The main points at issue concern (1) the degree of denuclearization, (2) the boundaries of the nuclear-weapon-free zone, (3) verification, and (4) the responsibilities of extra-zonal states.

1. A major controversy arose as to whether an undertaking by zonal states not to acquire nuclear weapons included non-acquisition of nuclear explosive devices for peaceful purposes. The problem was not new. It had been examined in detail during the negotiations which led to the signing of the Non-Proliferation Treaty (NPT) in 1968. It was then concluded that, for the purposes of non-proliferation, nuclear weapons and other nuclear explosive devices are synonymous and must be treated on exactly the same basis, because they contain the same nuclear components and require essentially the same technology. The states not accepting this conclusion have remained outside the treaty.

The Treaty of Tlatelolco settled the question with less finality, leaving an ambiguity. It allows explosions of nuclear devices for peaceful purposes 'including explosions which involve devices similar to those used in nuclear weapons', and spells out procedures for carrying them out (Article 18), but it also contains a reservation that such activities must be in accordance with its Articles 1 and 5: Article 1 prohibits the testing, use, manufacture, production or acquisition of nuclear weapons, while Article 5 defines a nuclear weapon as 'any device which is capable of releasing nuclear energy in an uncontrolled manner, and which has a group of characteristics that are appropriate for use for warlike purposes'. Some countries interpret these provisions as prohibiting the manufacture of nuclear explosive devices for peaceful purposes unless or until nuclear devices are developed which cannot be used as weapons, that is, practically for ever: the attached condition can hardly be fulfilled. Others dispute this view. The important problem of compatibility, or otherwise, of an indigenous development of nuclear explosive devices for peaceful purposes with participation in a nuclear-weapon-free zone agreement has remained unresolved.

The geographical extent

2. On the question of the geographical extent of the denuclearized area, objections were raised to a proposal for including in the zone portions of the high seas, straits used for international navigation and international air space. Again, the Treaty of Tlatelolco served as a frame of reference. The treaty defines its zone of application as embracing, upon fulfilment of certain specified requirements, not only the territory, the territorial sea, air space and any other space over which the zonal state exercises sovereignty 'in accordance with its own legislation' (some Latin American countries claim territorial waters as broad as 200 nautical miles), but also large areas of the high seas in the Atlantic and Pacific Oceans, hundreds of kilometres off the coasts of signatory states, over which no state at present claims sovereignty. In signing Additional Protocol II of the Treaty of Tlatelolco, France, the UK and the USA made it clear that they would not recognize any legislation which did not, in their view, comply with the relevant rules of international law, that is, the law of the sea. Their clarification amounts to a rejection of the postulate that the denuclearized status of the entire zone, as defined by the parties, should be respected. Also for the Soviet Union, an attempt to establish a high-sea sector subject to special status is unacceptable. This is one of the reasons why it has refused to join Additional Protocol II of the Treaty of Tlatelolco. Evidently, none of these nuclear powers is prepared to acquiesce in the limitation on the freedom of their navies to move or be stationed in international waters. To be valid, any such extension of the boundaries of a nuclear-weapon-free zone would need universal consent.

A divergence of views has emerged on the question of transit, that is, transport through the zone by carriers not belonging to a state party to the zonal agreement. It will be noted that the transit of nuclear weapons is not expressly forbidden under the Treaty of Tlatelolco. The Preparatory Commission of that treaty agreed that surface transit should be considered as excluded, while maritime transit, whenever allowed by a riparian state, must be subject to the provisions on the 'right of innocent passage' under the 1958 Geneva Convention on the Territorial Sea and the Contiguous Zone. The parties to the Tlatelolco Treaty contend that the nuclear-weapon powers which have undertaken, under Additional Protocol II of the treaty, to refrain from contributing 'in any way to the performance of acts involving a violation of the obligations of Article 1 of the Treaty', such as receipt, storage, installation or deployment of nuclear weapons within the zone, are expected to refrain from introducing nuclear weapons in the zone.

The USA and France, which hold the view that each party to a nuclear-weapon-free zone agreement should retain exclusive legal competence to grant or deny non-parties transit privileges, made a reservation to this effect upon signing Additional Protocol II of the Treaty of Tlatelolco. Other states, including the Soviet Union, maintain that all kinds of transit of nuclear weapons through the zone should be barred, including the entry into ports situated in the zone of vessels carrying nuclear weapons. (China has undertaken not to 'send its means of transportation and delivery carrying nuclear weapons to cross the territory, territorial sea or air space of Latin American countries'.) Their argument is that if nuclear weapons were allowed to transit the zone, even for a short time, the zone could not be considered as effectively denuclearized.

Control functions

3. There is a consensus that the obligation not to acquire nuclear weapons through manufacture or otherwise can be effectively verified, and that the central role in the control procedures should be given to the International Atomic Energy Agency (IAEA). But the experts are less specific as to the means of ensuring that the zone is free of nuclear weapons from outside sources. Such control functions would go beyond the statutory duties of the IAEA. An additional verification machinery has, therefore, been suggested. Its terms of reference would include inspecting naval vessels and military aircraft of nuclear-weapon powers within the zone, as well as checking whether nuclear weapons are not transported outside the zone by means of transportation belonging to zonal states.

4. While all extra-zonal states would be expected to commit themselves not to carry out any activity endangering the functioning of the zonal arrangements and, in particular, not to provide the states of the zone with any assistance that might lead to the development or production of nuclear weapons, the nuclear-weapon states would have to contract additional obligations, namely, not to deploy or stockpile nuclear weapons in the zone and, if they have already done so, to withdraw them. The undertaking to respect the denuclearized status of the zone should, in the opinion of most experts, include a formal pledge not to use, or threaten to use, nuclear weapons against any state included in the zone. However, the nuclear-weapon powers make such a pledge dependent on the content of each denuclearization agreement and have raised in this context the question of their participation in the negotiation of the zonal arrangements. They refuse to provide an unconditional assurance of non-use, even if all their postulates have been met. The USA and the UK would reserve the right to reconsider their obligations with regard to a nuclear-weapon-free zone state in the event of any act of aggression or armed attack by that party 'with the support or assistance' of a nuclear-weapon state. The USSR promises even less. It reserves the right to revoke its non-use commitment if a zonal state has committed aggression or has become an accomplice of aggression (irrespective of support or assistance by a nuclear power).

In view of the divergencies described above, Mexico proposed the acceptance of internationally valid definitions of the concept of a 'nuclear-weapon-free zone' and of the principal obligations of nuclear-weapon states. The proposal (with certain modifications) was later incorporated in a UN declaration adopted on December 11 1975 to the following effect.

> *Definition of the concept of a nuclear-weapon-free zone*
> A 'nuclear-weapon-free zone' shall, as a general rule, be deemed to be any zone, recognized as such by the United Nations General Assembly, which any group of States, in the free exercise of their sovereignty, has established by virtue of a treaty or convention whereby:
> (*a*) The statute of total absence of nuclear weapons to which the zone shall be subject, including the procedure for the delimitation of the zone, is defined;
> (*b*) An international system of verification and control is established to guarantee compliance with the obligations deriving from that statute.
>
> *Definition of the principal obligations of the nuclear-weapon States towards nuclear-weapon-free zones and towards the States included therein*
> In every case of a nuclear-weapon-free zone that has been recognized as such by the General Assembly, all nuclear-weapon States shall undertake or reaffirm, in a solemn international instrument having full legally binding force, such as a treaty, a convention or a protocol, the following obligations:
> (*a*) To respect in all its parts the statute of total absence of nuclear weapons defined in the treaty or convention which serves as the constitutive instrument of the zone;
> (*b*) To refrain from contributing in any way to the performance in the territories forming part of the zone of acts which involve a violation of the aforesaid treaty or convention;
> (*c*) To refrain from using or threatening to use nuclear weapons against the States included in the zone.

In other words, the countries deciding to conclude a treaty setting up a nuclear-weapon-free zone would themselves determine its provisions, including the extent of the denuclearization, the boundaries of the zone and the verification procedures. Once a nuclear-weapon-free zone has been recognized as such by the UN General Assembly, the nuclear-weapon states would be duty-bound formally to contract or reaffirm their obligations to respect the status of the zone and never to use nuclear weapons against a zonal state. France, the UK, the USA and the USSR were among a few dozen states that did not support the above declaration, and either abstained or voted against it. Experience has shown that General Assembly recommendations which are not unanimous, and which are opposed by states directly involved, are devoid of real significance. No state, and especially no nuclear-weapon power, is likely under the pressure of majority resolutions to alter a strategic doctrine which it perceives as vital for its security.

Conclusions

In and by itself the study has not helped to solve these controversial issues and it is doubtful whether it will actually enhance efforts to establish new zones in crucial areas of the world. Not only have the nuclear-weapon powers refused to recognize the status of zones established without their consent, or which would not fulfil the requirements they themselves have set for denuclearization, but they have made it clear that their possible non-use commitments could be withdrawn in case of war, whatever the weapons used by the aggressor. Thus, even within a nuclear-weapon-free zone a local conflict could assume nuclear proportions.

Many proposals for nuclear-weapon-free zones concern regions where countries have not yet renounced a nuclear-weapon option and have not joined the NPT; it would be unrealistic to expect them to do so under a more comprehensive arrangement. The elements of 'discrimination', about which they usually complain when referring to the NPT, would not disappear in a nuclear-weapon-free zone treaty. Besides, zonal agreements presuppose intergovernmental negotiations. It is difficult to envisage such negotiations where governments are unable or unwilling to communicate with each other. The establishment of a nuclear-weapon-free zone can hardly be a starting-point for peaceful relations among hitherto hostile nations. The reverse could perhaps prove true: the establishment of peaceful relations may be conducive to denuclearization.

12. *Disarmament through unilateral initiatives**

Recent developments in the field of armaments and disarmament are disturbing and should cause concern. The pace of armaments outstrips by far the pace of disarmament negotiations. While arms control and disarmament talks stagnated into protracted diplomacy, armaments revolutionized by modern technology have accelerated to such an extent as to bring in question the very modest results achieved in years of disarmament negotiations.

We now seem to have reached another critical turning point in the development and perfection of both nuclear and conventional weapons. Out of the testing grounds in Indochina and the Middle East came a 'quiet revolution' in conventional warfare. New generations of warplanes, warships and bombers were developed. Military laser technology with its ominous potential for destruction has seen the light of the day. A variety of new air and ground weapons, including 'smart' bombs, new types of guided missiles, an assortment of sensors and anti-personnel weapons were put into use. Electronics and computers came to dominate the so-called 'automated battlefield'. Following closely upon these developments, a mass of new conventional matériel is now proliferating to all continents, and Europe is no exception.

---

* Shortened version of a paper by Marek Thee from *Bulletin of Peace Proposals*, Vol. 5, No. 4, 1974.

On the other hand, vertical-qualitative break-throughs are reported in strategical weaponry. The notorious MIRVs, multiple independently targetable re-entry vessels, were joined by MARVs, manoeuvrable re-entry vessels, leading to a much improved accuracy in nuclear missiles. Simultaneously new kinds of 'miniature' nuclear arms, the so-called 'mini-nukes' combining lower explosive power with extreme precision of targeting have reached the production line. All these technological advances seem to have a strong impact on military postures. They tend to revise traditional war concepts and pose the problems of war and peace in a new way. Improved weaponry generated new scenarios for the conventional battlefield and led the military to devise new 'flexible' options for 'limited' nuclear warfare. Moreover, military experts were encouraged to speak about elimination of the threshold between nuclear and conventional warfare.

Technology and politics

It would seem evident that intensified efforts are needed to bring about a reversal of existing trends, and to infuse new dynamics into the disarmament process. As far as the Pugwash movement is concerned, one remark may be relevant: Pugwash has perhaps invested too much hope in technological solutions while the core problems seem to be political. Whatever theory one applies for the explanation of actual armament dynamics—the action–reaction model, the autistic or internal paradigm (spelled out by SIPRI as technological, economic and bureaucratic pressures), or the pure political interpretation pointing on the one hand to the role of imperialism and on the other communist expansionism—the political element seems predominant. Indeed, political wisdom and will are needed to face and stand against the powerful combined pressures of research and development, the vested industrial interests, the military establishments and the cold war ideological dogma. Thus, while no efforts should be spared to urge change and to press on with current arms control and disarmament negotiations, new and unorthodox departures must be sought for a fresh take-off in the reduction and limitation of armaments.

Unilateral reciprocated initiatives

One of the strategies which may have a salutary effect on disarmament is de-escalation based on unilateral initiatives. Started by exemplary unilateral cuts in armaments by one party, the de-escalation process could through reciprocation develop into a chain reaction of graduated arms reduction. Initiative and challenge would unite to generate a new environment and new pressures for disarmament. Once initiated, no party could easily drop out of the process, which could achieve a snowballing effect. Of importance is the fact that successive unilateral initiatives could be graduated in risk so as not to jeopardize one's own basic security. Indeed, because of the diversity of the strategic scene, its quantitative and qualitative aspects, its dispersion in space and time, and the multiple functional elements in the military build-up, opportunities for unilateral initiatives with no damage to real security may be very numerous. One could, e.g., start with purely confidence-building measures, with partial freezing of military budgets, with greater openness in military debate, with reduction of some offensive in favour of defensive weapons, or—what might be of overriding weight—with a gradual diminution of military research and development. From the point of view of immediate application, the talks on mutual force reduction in Europe may well offer good openings for action.

Of course, a primary condition for the success of such a strategy is reciprocation. Each unilateral step by one party would have to be followed by similar actions by the other party. But considering the existing political dynamics and, in effect, the changed atmosphere of détente, one cannot possibly see how such a challenge could be left unanswered. On the other hand, should the strategy work, one might arrive at an incremental cumulative process leading progressively from simple partial moves to wide disarmament measures. Unilaterality combined with gradualism offers good prospects for enhancing confidence, reducing fear, and leading to dismantlement by degrees of the military build-up.

Favourable conditions

The idea of gradual unilateral disarmament moves is not new and has been debated in circles concerned with disarmament for quite a long time. However, objective conditions, a perceived assymetry in the state of armaments in East and West have made its application difficult. Taking into consideration the perceived superiority of the West in resources and technology, protagonists of unilateral disarmament initiatives tended to direct their appeals almost exclusively to the United States. But US strategists and political leaders were not willing to listen.

Yet in recent years the situation underwent a basic change. The balance of forces has been transformed to a state of parity and symmetry in actual military strength. This then opens new perspectives and new possibilities for reciprocated unilateral disarmament initiatives. The way is now open for both sides to take the initiative. The United States may in new conditions be willing to reconsider its resistance to unilateral moves, and the Soviet Union might be willing to come forward with moves liable to trigger a chain reaction of real disarmament. In fact, political considerations and military interests linked to the new strategical constellation between East and West tend to favour a new approach to de-escalation in the arms race.

The political advantages

From the point of view of concerned involvement in the peace movement, the strategy of graduated and reciprocated arms reduction initiatives offers some exceptional advantages.

First, this strategy presents a unique opportunity to activate public opinion and solicit its support in favour of arms control and disarmament. It is no secret that after some animation and engagement of a fairly broad section of public opinion in support of disarmament negotiations at their initial stage, the momentum was later lost. Both the turn to technicalities in the negotiations and the protraction of the process had an impact on the movement. Additionally, détente may have contributed to calm anxieties and reassure minds. In reality, people are largely confused and feel alienated from the diplomatic process. The grasp of the issues involved has receded. Unilaterality combined with reciprocation would invite greater curiosity on the part of the public and would offer larger opportunities for action. Clear goals could be set for pressuring issues in scope and time, and concrete roles could be devised both for mass movements and for specific professional milieus. Obviously, scientists and scholars standing in the centre of the technological revolution could assume special responsibilities.

Secondly, the strategy of unilateral reciprocated disarmament initiatives by definition tends to alter the very environment and climate of mutual interaction by opposing parties. It is a strategy with inherent confidence-building dynamics. One of its most important aspects is that it tends to remove secrecy and to bring the negotiations into the open. Surely, an atmosphere of openness would favour objectivity and a more constructive approach to disarmament. Both establishments and individual negotiators would have to be more accountable and face up to their responsibilities more candidly. Thus, unilateral initiatives would sustain those forces sincerely striving to reduce tension, and would tend to counter manipulations aimed at using protracted negotiations as a screen for actual armaments.

In summary, unilateral reciprocated disarmament initiatives would respond to the long felt need for making the disarmament process more transparent and understandable to public opinion. Disarmament talks and measures would acquire greater concreteness, and drawing strength from the climate of détente would themselves contribute to a better atmosphere of understanding. The process of disarmament could be accelerated and further armaments halted. Unilateral moves may not solve all the problems. But as an alternative and supplement to current endeavours they are full of promise. They may be a step in the right direction.

*13. Disarmament versus arms control**

Throughout this century, disarmament and the control of armaments have been more or less permanent features of the foreign policies of the great powers. And since World War II, literally hundreds of international

* From *Nuclear Disarmament or Nuclear War?*, SIPRI, 1975.

meetings have taken place mainly in an attempt to control the nuclear arms race between the USA and the USSR and then to reduce the number of nuclear weapons in the arsenals of the nuclear-weapons powers. But, in spite of so much talk, virtually no progress has been made towards either of these ends. How can this be explained?

A major reason for a lack of progress towards nuclear disarmament is related to the decision taken by politicians in the early 1960's to abandon attempts at the direct negotiation of general and complete disarmament and to work instead for partial arms control measures. The idea was that, by this method, it would be possible to move towards general disarmament by small steps.

Whereas 'disarmament' normally means a quantitative reduction in total numbers of existing weapons by the traditional methods of international negotiation, leading to a multilateral treaty, 'arms control' normally refers to negotiated measures leading to the slowing down (and eventual halting) of arms races. In other words, disarmament refers to the elimination of armaments (either specific armaments, such as nuclear ones in the case of nuclear disarmament, or all armaments in the case of general and complete disarmament) and arms control to curbs on acquiring new weapons.

Slow progress seems to be inherent in the arms control approach. Arms control advocates claim that, in a world of security-conscious sovereign states, disarmament can only come about—if at all—as the end product of a lengthy process. The initial stage in this process involves banning weapons of little or no military value (such as biological weapons) and banning weapons from environments of little or no military significance (such as the sea-bed, outer-space, Antarctica and so on). This process, so it is said, will establish such a degree of mutual confidence between the powers and so improve the climate of international affairs that, in due course, far-reaching disarmament may be possible.

Confidence-building measures—like arrangements to minimize the risk of 'nonprovocative' weapons, the adoption of 'nonprovocative' strategies, and so on—and tension-reducing measures—like nuclear-free zones, demilitarized areas, nuclear test bans, nonaggression pacts and so on—are advocated as measures which may facilitate the eventual negotiation of disarmament. And, in the longer term, it is suggested that international peace-keeping institutions can be established, mechanisms for crisis management can be elaborated, verification systems can be worked out, and so on. These, the arms controllers insist, are essential prerequisites for any general disarmament negotiations rather than measures to be incorporated into a disarmament programme or established parallel with the implementation of one.

As might be expected, there is some common ground between the advocates of arms control and those of disarmament (whether nuclear or general). The first phase of a typical disarmament plan would, in fact, include a number of arms control measures. But there is a major difference between the two groups on the question of timing. Apart from the danger of irrational behaviour and accidental nuclear war, disarmers point to the extremely rapid advances taking place in military technology as a reason for urgency. Weapons are being developed and deployed which produce periods of considerable instability in international affairs. It is true that periods of rough parity occur between the strategic nuclear forces of the two great powers, but then new weapons emerge which significantly upset the balance. Sooner or later, a period of instability may occur, for example at a time of severe international tension and also possibly when one or more of the states involved is led by an irresponsible leader. Disarmament advocates see extreme dangers in such a combination of events. They also feel that these dangers will be multiplied by advances in nuclear-weapon technology (such as the improved accuracy of warhead delivery) and if many more nuclear-weapon powers emerge.

Although disarmament advocates disagree amongst themselves on the details of the most desirable general and complete disarmament plan, most agree on the urgent need for nuclear disarmament—usually as part of some comprehensive programme of disarmament. Arms control advocates often argue that disarmers exaggerate the dangers in a nuclear-armed world and, more importantly, they feel that it is politically unrealistic to expect significant disarmament to be achieved in a world organized as ours is today. In their turn, disarmers argue that far-reaching disarmament is a politically realistic objective (although admittedly a difficult one to achieve) in a world of sovereign states because governments could be persuaded to disarm by, for example, the pressure of public opinion. They would then direct their energies towards disarmament instead of diverting their attention to the negotiation of partial arms control measures.

Misconception of the public

The arms control efforts—multilateral and bilateral—over the past 15 years have failed to produce any nuclear disarmament or even to halt the nuclear arms race between the USA and USSR. In spite of this, the public has been seriously misled into believing that steady progress is being made in disarmament. The main reason for this mistaken belief is that political leaders habitually make euphoric statements about the value of arms control treaties. Each treaty is signed with much pomp and ceremony, and to the accompaniment of speeches full of high-sounding promises of bigger and better things to come. And the preambles and articles of the treaties usually contain far-reaching commitments to further progress which are rarely, if ever, followed up.

The political leaders, are, of course, aware of the almost universal human desire for a secure and peaceful disarmed world and of the considerable political benefit to be gained from paying lip-service to this desire. But perhaps not surprisingly they, and those involved in negotiating arms control treaties, take on a 'professional optimism' which apparently causes them to convince themselves that substantial progress is being made. This may be a psychological necessity for those involved but it also hampers progress toward disarmament.

Time and energy spent on negotiating partial measures are time and energy diverted from negotiating real disarmament. The present state of world armaments is such that in this latter task we have precious little time to lose.

*14. International control of disarmament**

Without any doubt, the request for tight control, on the one hand, and the stubborn unwillingness to accept verification by direct means as inspection, on the other hand, have been a stumbling block in the disarmament negotiations up till now. The United States has from the beginning been pressing for far-reaching controls as a pre-condition for its preparedness to enter into agreements of any military significance. The Soviet Union, on its part, has equally consistently declined to accept inspections within its territory when any agreement has been under active negotiation; for the ideal state of a totally disarmed world the Soviet Union also was prepared to accept quite a profusion of inspections.

Behind, and under, their outwardly often fierce disagreements, inducing the non-aligned powers to use their ingenuity to construct and propose compromises, there has all the time been a secret and undeclared collusion between the superpowers that neither of them wanted to be restrained by any effective disarmament measures.

The need for controls

The purpose of controls on disarmament agreements must be to have assurance that they will not secretly be broken. The historical record, however, does confirm to the contrary that clandestine violations of agreements are not often committed. In the armaments field they may hardly ever have occurred. Open breaches are rather more likely than clandestine ones. On the other hand, if weapons, their deployment and production have become more difficult to observe from outside, the means of detection have become vastly more effective.

Not only has the organized spying, on both sides, been kept up but many new means of accurate monitoring from a distance have been added, without requiring inspections for verification of what goes on in other countries. They have been by-products of the arms race. Satellites are the prime example, others are radar and acoustic devices like sonars for marine surveillance, further the examination of emissions, e.g. of radioactivity or chemical substances, the analysis of telecommunication patterns, and the list could go on. Regrettably, the most effective of these monitoring devices are now monopolized by the two superpowers, above all the satellite

* Excerpts from Alva Myrdal: *The Game of Disarmament*, Stockholm 1976.

systems. So much more unreasonable is it that it should be on their resistance, often just on the issue of controls, that the disarmament negotiations are floundering or resulting in less than effective treaties.

When disarmament negotiations were about to start more in earnest in the beginning of the 1960's, ambitions were fixed at a very high level, aiming at a state of world affairs, where individual nations were disarmed and the United Nations was left the duty of collectively policing the world society. In preparation for the work about to begin in the Geneva Disarmament Committee, the superpowers reached agreement on a set of important guidelines for the negotiations on General and Complete Disarmament. They are generally referred to under the name of the McCloy-Zorin Agreed Principles, which were also approved by the 1961 UN General Assembly. The sixth of the agreed principles reads:

> All disarmament measures should be implemented from beginning to end under such strict and effective international control as would provide firm assurance that all parties are honoring their obligations. During and after the implementation of general and complete disarmament, the most thorough control should be exercised, the nature and extent of such control depending on the requirements for verification of the disarmament measures being carried out in each stage. To implement control over the inspection including all parties to the agreement should be created within the framework of the United Nations. This international disarmament organization and its inspectors should be assured unrestricted access without veto to all places as necessary for the purpose of effective verification.

However, much attention to public documents and learned treaties on disarmament issues have devoted to the issues of controls, it has not lead to any commonly shared clarification about what is the *function* of controls, what *sort* and *degree* of controls are *needed*, *when* controls should set in, *what* should be controlled and *by whom* and by what *methods*.

Ex post and ex ante

In order to sort out the vexatious issues of control and verification, the overriding question should first be answered: what is the function of verification in relation to disarmament treaties? I would forcefully argue that the main function of control rules is to deter from violations of an agreement, not to track down violators in order to prosecute them. Thus, assuming that it is desirable to be able to control faithful observations of obligations, the practical question is how we should construct verification methods which are applied from the outset, and so clearly and openly and as far as possible automatically, that they serve as a constant warning to would-be violators that clandestine adventures will not succeed. I would call this the *ex ante* approach. Hitherto, far too great interest has been placed on finding verification methods which could be utilized by one or several parties to a treaty after a suspicion has arisen to secure evidence of a possible violation of treaty obligations and on this ground bring the culprit to court. This I would call the *ex post* approach.

Starting from a theoretically different angle controls should have the very important function to prevent violations, thus placing deterrence ahead of detection. In an *ex ante*, before-the-event approach, such methods of control and verification must be utilized, that sufficient risks of disclosure act as a deterrence against any Party attempting to violate a treaty obligation. In the case of the test-ban issue such methods exist in the highly developed system of current and continuous observations from seismic stations and satellites.

The type of control and verification which relies on police surveillance and aims at a verdict by judicial authority must be avoided as far as possible also for the sake of not poisoning the international climate. Instead, a gradual process of mutual contributions by the parties concerned to clarify matters would engender trust. The Swedish delegation has in the international disarmament negotiations proposed such a system of a largely voluntary control procedure, called 'verification by challenge', relying on the interest of a party under suspicion 'to free itself through the supply of relevant information, not excluding an invitation to inspection by an outside party or organ', as I briefly described the concept in 1970.

In more concrete terms it may be said that in this scheme of verification by challenge, or, as it is also known, 'challenge and response', the *sanction of publicity* is relied upon as one of considerable efficiency for deterrence.

Openness is the primordial tool to be used in verification of disarmament. But to be practical and acceptable as a demand directed to governments and national authorities it can not be categorical. It must be seen as a request for more open accountability, moving along a scale, from an outer circle of uncompromising openness towards an inner circle where the military interests will buttress secretiveness to the last. Accessible to verification by the international community must first be the facts at the outer rim as disclosed by scientific and technological R&D, to be completely made available by publications and other information media. In a middle field, gradually made to expand its open sectors, belong facts as revealed by national budgets, trade and production statistics, satellite observations about both civilian and military resources in other countries. And so on. Further in, military interests will more avidly shield information about weapon strength, deployment of forces and similar arrangements of immediate concern, although some encroachments on the policy of secrecy are now being favored, for instance, by the agreement at the European Security Conference about notification of manoeuvres.

Reducing secrecy

It can be safely stated that for both the *ex post* verification and the *ex ante* deterrence to function with maximal efficiency, a main precondition would be an increase of the scope of knowledge about what happens to armaments as well as disarmament in the several countries. It would be particularly true that the more pertinent knowledge that were available, the more reliance could be placed on the *ex ante*, the before-the-event approach to verification.

Towards the end of World War II Niels Bohr suggested that the way to deal with the problem of nuclear arms was none but the radical one of abandoning the secrecy surrounding the technology of nuclear explosions.

In regard to national policies, openness is not enough. It must be coupled with an honest desire in every country to help the citizenry effectively to share the knowledge. More openness should be reassuring within a nation and between nations. I would claim it as the great freeway to disarmament control. It is also in line with the cultural trends of our time, as information is flowing in ever fuller streams within and between our societies. The masterkey to the whole problem of control of disarmament is not reliance on specific formulae in a treaty, but the construction of a firm basis of universal confidence through a cumulative process of fully shared factual information.

International verification of disarmament?

The history of disarmament negotiations is punctuated by proposals that there should be created an international agency for controlling disarmament. The McCloy-Zorin Agreed Principles presurmised that for implementation of disarmament measures within a foreseen treaty on General and Complete Disarmament, such measures should 'from beginning to end' be under 'strict and effective international control' through 'an international disarmament organization'.

In the painfully protracted but unproductive negotiations on the comprehensive treaty drafts, which followed for some years in the Geneva Disarmament Committee, not much attention was devoted to this organizational matter. And then the whole pre-occupation with plans for General and Complete Disarmament faded away.

Four major questions *vis-a-vis* any international verification organ have to be resolved. They concern (*a*) structure and powers; (*b*) scope of functions; (*c*) status as independent or attached to some existing organ; and (*d*) establishment date.

The overriding question as to structure of a verification agency is whether it should be streamlined only for one treaty or made to serve several. An innovation, of truly creative importance, would be the central role of the review conferences, as this presupposes doing away with the irrational reliance on 'depositary governments', which has been allowed to mean new nuclear weapon powers. Having the International Verification Agency to call review conferences for checking on and possibly amending various disarmaments agreements would create a pattern of one verification agency servicing different constituencies, as these legally competent bodies are varyingly composed by Parties to individual treaties. It may be noted that also regional treaties might come to utilize the services of the agency as a joint secretariat.

In regard to the scope of functions there are also important issues to discuss. As of today the non-governmental but international SIPRI as part of its work fulfills most of the pre-verification needs. As of tomorrow it is possible that the UN Secretariat will get expanded resources to take on some of the valuable documentation tasks. But the central work of verification, politically sensitive as it is though all-important for the credibility of disarmament agreements, can never be placed directly under a UN organ, be it the Secretary-General or the Security Council. It required a prestige of scientifically based objectivity and thus a status of fearless independence. Most emphatically should be underlined that the role of the Security Council is only at the top as judge on possible violations, not at any intermediate station. The separation of the investigative and the jurisdictional functions, referring them to different organs, must be made clear and explicit.

The status of an International Verification Agency depends on two conditions *sine qua non*: that it is intergovernmental, and that it is independent. While strongly emphasizing the independent character, but taking into consideration a number of practical desiderata, I would suggest an intermediate position, not one as independent as that of the Specialized Agencies, like WHO, but neither one totally dependent on the central organs of the United Nations. The best parallel to seek is probably the semi-independent status of the UN Environmental Program (UNEP). A relatively high degree of independence is singularly needed for an organ whose whole respectability, yes, usefulness, hinges on its objectivity and freedom from political shackles. Like the IAEA it should, however, report directly to the UN General Assembly, according to a resolution under its own item, not having any reason like UNEP to channel its report through the UN Economic and Social Council.

A decision should be taken at the earliest possible date, i.e. after the completion of the preparatory study, to set up the agency immediately, on an interim basis. The main reason for this haste is that the agency should function as an instrument to accelerate the negotiations on agreements, taking over and effectivizing greatly the valuable work so far done by *ad hoc* meetings of experts, sometimes organized by the Geneva Disarmament Committee, although too few and too far between, and without any organized follow-up work.

There is no need for any waiting time. There would certainly be a quorum available for a majority decision in the UN General Assembly for a resolution to start on the first two steps: the immediately implementable one of a study and planning group, and the agreement in principle to set up an International Verification Agency on an interim basis.

Some encouraging steps towards greater international participation in space activities, at least in regard to the use of earth sensing and broadcasting satellites, are now being made in the United Nations. Its Committee on the Peaceful Uses of Outer Space has in 1975 recommended that action be taken towards a UN Satellite System to service member nations. So far, these plans envisage earth resource studies, e.g. for crop inventory, but they do imply some encroachments on the near-monopolies of the superpowers over space activities.

International access to data from satellite monitoring will come to be an absolute necessity for a truly serious work of verification of disarmament agreements—it would also be valuable for 'early warnings' on changes in the world's armament picture. While it may sound overambitious to raise these problems at this juncture, it must be foreseen that the world community should not tolerate to remain forever the poor relative when it comes to safeguarding international peace and security.

When so much attention has here been bestowed on an international verification agency, starting with chemical disarmament, it is because the time is ripe for a bold initiative at the international level. Rightly, however, what is to be envisaged is not so much an agency for verification purposes as a system, built on national as well as international responsibilities. Even when I advocated the establishment of a verification system over a wider vista of disarmament measures, I underlined that it must consist in a comprehensive scheme organically and hierachically built up from the national to the international levels.

There is evidently a growing acceptance of the idea of national controls, coupled with duties to report to an international body. There should be no doubt that if only an agency, at the international level, is given the capacity to handle a data exchange, there would be a spate of information forthcoming. But, and this should be stated as a praeterea censeo: it is urgent also to propagate the thesis that monitoring activities and even institutional arrangements for verification and surveying of progress in the direction of arms regulation measures can proceed even in the absence of formal treaties. More than that, they might well, by beginning to prove their worth, stimulate the emergence of formal disarmament agreements, of which many are so long overdue.

Internationalizing knowledge

All kinds of positive interests converge in a claim for widening the margins of knowledge and internationalizing the access to and distribution of relevant information. I have here stressed what it would mean for fortifying deterrence against cheating on disarmament agreements. In addition, there are two important groups of customers who could reap very great benefits by a reduction of military secretiveness. One is the scientifically and technologically less advanced countries. The other is the academic world republic of scientific workers.

The interests in internationalizing knowledge and utilization of new technologies for development purposes should be made to focus much more sharply than hitherto on the advantages which are being derived from advances in military R&D and which now accrue practically totally to the already highly developed countries. Scientific research on the whole is to some 98 per cent pursued in laboratories and universities of the rich countries. It predominately serves their interests. In turn, about half of that work for crossing the frontiers of knowledge is devoted to military research.

Nuclear science, electronics, aeronautics, materials research, particularly in the fine structure of metals, data processing by computers, solid-state physics, hydrodynamics, oceanography and space science are examples of remarkable recent progress that has burgeoned under the impetus of military research after the war.

This has taken place simultaneously with the new era of decolonization that opened our eyes to the needs of the Third World. And these needs for development warrant no restatement here, nor the connection between their economic underdevelopment with the underdevelopment of science and technology. It should go without saying that access to all knowledge, and not least the most advanced, must be internationalized.

As the research and development resources of the world are distributed, and utilized, they constitute a seemingly built-in assurance that mankind will forever be divided in a have and a have-not category of nations, and that the detrimental effects on the weaker will, in relative terms at least, be growing with each year. Therefore, the most natural of all supportive partners who should join up with the disarmers ought to be the assembly of the claimants for development within a new economic world order. The open release of information should be one of their major demands.

Another section of interest which is harmed by the present lack of freedom of information is Science itself. The cross-fertilization of ideas, so necessary for progress, should not so often need to be stopped by bureacrats at national or even agency boundaries. Particularly the secretiveness dictated by military considerations is holding down much scientific work to an artificially and unnecessarily unproductive level. It obviously creates a personal and professional dilemma that is experienced with varying strength by individual scientists, when national agencies use their discretionary power to keep 'classified' and hidden scientific results and to stifle the natural flow of information.

Attempts have been made to organize scientists for a defense of their freedom of research, sensing that they are being misused to work for destructive rather than constructive purposes. Codes of ethics have from time to time been suggested. So, for instance, has the Pugwash movement tried to lay down rules parallel to the Hippocratic oath for doctors. At its 22nd conference in Oxford in 1972 a working group dealt with the problem area of scientists and society. The group recommended for the consideration of the Continuing Committee the evolution of a 'Pledge for Scientists', based on the following text which is similar to that considered by the Royal Norwegian Academy of Sciences:

> I will not use my scientific training for any purpose which I believe is intended to harm human beings.
>
> I shall in my work strive for peace, justice, and the betterment of the human condition.

This is not the place to dictate a new morality for scientists in the various fields. But we should all, as citizens of the world, be aware of the need to free them as much as possible from the shackles of military secretiveness.

15. *Crisis in arms control**

From 'détente' to 'peace through strength'

There are signs that efforts at arms control are in deep crisis. The symptoms are many.

Firstly, arms control negotiations have reached a stalemate, bogged down by political and technical controversy. After a decade of relative success in which a number of agreements were produced, arms control can neither be said to fulfil the promises inscribed in the agreements concluded nor to advance to new meaningful accords.

Secondly, parallel with the deadlock in arms control efforts and despite accords concluded, the arms race continues unabated with a constant upward rush in armaments, nuclear and conventional, quantitative and qualitative. There is acceleration and escalation in the arms race. In a world plagued by economic recession and diminution of raw materials, armaments flourish as never before. The arms trade has exploded. And acquiring new dimensions the arms race has encompassed the whole world: the superpowers and small nations, the rich and the poor, East and West, North and South. New weapons have become a status and prestige symbol as well as a prime instrument in the diplomatic game. They actually kill daily hundreds of people in scattered local conflicts around the globe with a cumulative toll of tens of millions since World War II.

Thirdly, signs of the crisis are also evident in the political domain. There is strain in the climate of international relations, especially between the main powers engaged in arms control talks. This has found expression in, among other things, the rejection by President Ford of the term *détente* and its substitution by the slogan of 'peace through strength'. True, there has always been some confusion about the concept of *détente*. Yet whatever the relaxation of tension brought by *détente*, its general interpretation has assumed an evolution which would complement political gestures by military deeds, political *détente* by military détente—an evolution towards actual disarmament. But, as admitted by Secretary General L. Brezhnev at the 25th Congress of the Communist Party of the Soviet Union, 'the arms race is becoming more intensive'.

Arms control at its limits?

What is the nature of the actual crisis in arms control? Is it only a transient phenomenon? Or, are the troubles of a more fundamental character?

Looking back on past experience of arms control, to its origins and performance, it is safe to say that there has always been some basic weakness

* Shortened version of a paper by Marek Thee from *Bulletin of Peace Proposals*, Vol. 7, No. 2, 1976.

in the concept of arms control. At its inception, arms control offered a 'realistic' scheme for regulating the competition in armaments, in lieu of the 'utopian' goal of disarmament. The design was to substitute partial measures for comprehensive measures and try to steer armaments by accord. Arms control intended to 'balance' force levels and adjust armaments in such a way as to reduce the probability of war, rationalize arms spending and keep developments under some control even in the case of war. But apart from the very material fact that disarmament was dropped as an immediate goal, there was also an innate contradiction in this endeavour: the gap between means and ends. As experience has shown, it proved in fact impossible to master the highly complex and dynamic armaments environment by limited piecemeal measures such as the Partial Test Ban Treaty or the vague Nuclear Non-Proliferation Treaty, not to speak of pure declaratory moves or nuclear non-armament agreements in regions outside the real arms race such as the Antarctic or Outer Space.

The aims of arms control in the context of SALT were redefined recently by Secretary of State Kissinger in his interview with the *U.S. News & World Report* (March 15 1976). The goal, according to Kissinger, is: (*a*) to make it 'less likely for either side to achieve a decisive advantage in strategic weaponry', (*b*) 'to insure that these weapons will be used only in the most extraordinary circumstances', and (*c*) in case of war 'nonnuclear means would always be preferable, but I don't want to exclude nuclear means in certain situations'.

Such an approach, naturally, is highly unsatisfactory and of no relevance to disarmament. Its structure rests entirely on the 'realistic' acceptance of an international threat system. Instead of arms reduction, the proposition is of constant preparation for war.

In actual performance, the weakness and fragility of arms control have been brought into strong relief.

Firstly, arms control has not been able to cope with the rapid technological advancement in armaments. As the centre of gravity of the arms race in recent years has moved from quantity to quality, arms control found it more and more difficult to devise ways and means to keep soaring armaments under control. Technological innovation has tended to outstrip the pace of negotiations. Moreover, the protracted negotiations have rather favoured the development of new arms, first as a bargaining chip in the initial stage of talks and later as a trump card in the balancing exercise. In fact, freedom of arms modernization has been expressly inscribed in the SALT agreements. Underlying arms control paralysis to tackle problems of technological advancements in armaments has been its inability to interfere with the powerful military research and development establishment. Not only are the innovative pressures of military research and development sustained by a variety of vested political, economic and military interests, but the very mode of operation of military R&D generates autistic dynamics which tends to escape control. New technology has revolutionized the armaments process and perpetuated the arms race by the so-called follow-on imperative involving long cycles of design, production and deployment of new arms. The long lead times of military R&D tend to upset calculations based on existing arms. Thus, unless military R&D is included in disarmament designs, any effort at arms control will always be subverted by the momentum of constant refinement of existing weapons, development of new weapons and invention of new weapon systems. New generations of arms, more sophisticated and destructive, naturally tend to erode agreements dealing with weapons turned obsolete.

Secondly, efforts to balance force levels of the superpowers—to arrive at 'essential strategical equality'—as assumed by arms control have proved to be an almost futile exercise. The very dimension of military 'equivalence', vague and ambiguous as a term, is hardly measurable. This is so especially in conditions of large asymmetries in force structures caused by different national security requirements, geopolitical exigencies, divergent strategical doctrines and varied perceptions of defence and offence. And on the top of all comes the secrecy prevailing in military affairs which precludes full assessment of actual force composition and level of strength of the contending parties. The larger the military build-up and the more intense the technical innovation, the more complex all the above elements of the situation. The more unmanageable also the task to balance the overall military strength. The real outcome in the arms control dealings of recent years has been a steady race and subsequent mutual acknowledgement of ever higher and more costly levels of 'equivalence'.

Thirdly, the concept of arms control born in conditions of a bipolar world military order and tailored to negotiations in a military environment dominated by the United States and the Soviet Union is losing its relevance in a world radically changed in recent years. Arms control resists adaptation to a multipolar world constellation. Fragmentation in the international system and the spread of nuclear technology as well as most sophisticated weapons to all corners of the world has made arms control a much more complex proposition. Developments have become less predictable and the need to accommodate a number of divergent interests has heightened the challenge. In the process, a clear contradiction has become evident: while arms control in the bipolar context is basically *status quo* oriented, transformation on the international scene has required change. Also, the main doctrinal pillar of arms control—the theory of deterrence with its first and second strike scenarios—is becoming less applicable in a varied multipolar context. Strategical calculations have acquired new dimensions and the assumption of rational behaviour of leaders is becoming at least questionable much more than before. Would arms control be able to adapt to the new circumstances?

Fourthly, in the changed political and technological environment the very negotiation process within the framework of arms control becomes more and more intractable. One of the reasons lies in the fluidity of the situation and especially in the accelerated arms race caused by an accumulation of action–reaction and autistic dynamics. While talk in the negotiations is of 'parity' and 'equivalence' of forces, the steady race in armaments generates suspicions of a rush to 'superiority'. Confidence is constantly being undermined, and compliance with agreed provisions is becoming again and again an element of controversy. But the search for tighter and more controllable accords clashes with the dynamics of technological innovation and the wide spectrum of armaments, strategical and tactical. There is thus a crossing of substance and procedure which tends to check the machinery of negotiations and arms control.

The search for alternatives

Whatever the ultimate conclusions from the above analysis, there is no doubt that arms control groups in order to survive, must seriously rethink the issues involved and search for a way out of the deadlock. A new momentum for disarmament is urgently needed. Partial technical solutions which only touch the problems on the surface and do not reach to the structural roots are obviously insufficient. They have outlived their purpose.

We have come to realize that despite a number of arms control agreements and despite the appearances of *détente*, international security has not been strengthened. On the contrary, the powder keg we are sitting on has grown immensely. The rule which governs events is: the more modern weapons and the greater their sophistication, the greater the vulnerability of all parties, big powers and small nations. Efforts to stabilize the situation by piecemeal arms limitation steps have had little effect and the danger of destabilization by non-synchronized technological breakthroughs is constantly with us. Any spark of crisis, regional or central, can ignite a fire. There is a kind of quagmire dynamics in the present arms race: the more we reach out for new arms, the deeper we sink in insecurity and conflictual postures. And, it should be stressed, most of the modern arms produced in recent years, with their emphasis on accuracy, manoeuvreability and penetrability are more offensive than defensive in nature.

Security is a complex process. It relies today on the triangle binding together political, military and technological components. To move things on the military corner it seems now imperative to master developments on the political and technological corners—to activate political will at the centers of power and bring military research and development under control. Accordingly three tasks stand out in the field of arms control and disarmament: (1) to arrest the cancerous growth of military R&D and divert its creative potential to human needs, (2) to introduce greater openness in security planning and arrrangements so as to make issues more transparent and generate greater confidence in disarmament negotiations, and (3) to democratize the very process of negotiations so as to involve all key factors and international public opinion in shaping security for all.

As to direction of change, a strong case can and should be made for a return to the idea of comprehensive measures within the framework of general and complete disarmament. Arms control agreements together with the rhetoric of *détente* have created the impression that we have successfully set foot on the road towards disarmament. Some seem even to believe that real progress in disarmament has been made. In reality, however, the whole arms control exercise is about how to co-ordinate and guide armaments.

With arms control bogged down in an impasse, it is time to search for alternatives. The world greeted with high expectations the 1961 Soviet–American understanding on Agreed Principles for Disarmament Negotiations, which provided for general and complete disarmament. It seemed then a workable proposition despite contemporary cold war tensions. The promise of general and complete disarmament has subsequently been inscribed in almost all arms control agreements of recent years. All the more reason it should serve as a realistic approach today in an atmosphere of reduced tension. As arms control realism has proved to be a utopia, perhaps the 'utopia' of disarmament may turn out to be the most realistic way to take. It is certainly the only realistic option considering the fact that the very survival of mankind is at stake. We desperately need real progress in disarmament. The arms controllers community and all people concerned with the security of the globe face a momentous challenge.

16. *Agenda for disarmament**

Against the background of the long history of defeats to reach effective agreements on important disarmament measures, it might look as if the task were impossible. But I decline to accept a defeatist view. The arms race is irrational and we are not permitted to let unreason stand unchallenged. The hope to reverse the trend must be built on untiring efforts to educate peoples and governments to see their true interests. This must mean effectively attacking the forces within the countries which are propelling 'the arms race within the arms race'.

They all converge in what Dwight Eisenhower in his last presidential message to the American nation called the 'military-industrial complex', to which 'academic' must be added. These forces are represented in legislative assemblies. They are probably particularly strong and visible in the United States, but they operate in all countries, not excluding the Soviet Union which so often has demonstrated that it feels itself in fierce competition with the United States. As forces driving the arms race forward they are irrational not only from the international but also directly from the national point of view of the individual states, including in the first place the superpowers.

Besides opposing the arms race on account of its irrationality we must also do so on the basis of simple morality. With each new generation of them, arms become more inhumane. The tendency is that every new war

* Excerpts from Alva Myrdal: *The Game of Disarmament*, Stockholm 1976.

becomes more cruel. Since decades and even centuries certain rules about how warefare is not to be conducted have been accepted as international law, but they are now disobeyed in the most flagrant way. The rules must be expanded, modernized and respected. They must begin to guide efforts to restrain not only the use but also the production and trade in cruel weapons.

The sober conclusion we, who stand for the ideal of proceeding towards disarmament, have to draw is that in the present world situation what we must aim at is a strategy for limitation and reduction as part of a comprehensive programme of specific arms and other preparations for war. On the other hand, attempts hitherto made to agree on some such specified arms limitation measures have been far too piece-meal. No prospectively important treaties have been concluded to abstain from production and use of weapons, only some marginal fringes being tied up, and even that not effectively. On the really important issues, and in the first instance the nuclear arms race between the superpowers, it is only fair to say that the negotiations up to now have been futile.

There must be found a strategy for limiting, reducing arms and eliminating some of them, that is both realistic and vastly more courageous in regard to all the several issues. A new comprehensive grasp of the disarmament problems, while abandoning the attempt to get a general agreement in one stroke, must seek a solution that in a perspective way is compounded by specific but integrated measures. The measures as such may take the form of international legislation, multi- or bilateral agreements or just unilateral moves forward on the road to the acknowledged goal.

Nuclear weapons

Our future is being more heavily mortgaged by the acceleration in the qualitative arms race. That is now the superpowers' master game, underpinned by their near-monopoly on ultra-technologies and turning them towards new strategies of first-strike superiority. It is therefore imperative that a barrier be clamped down as immediately as possible against further competition about improvements of weapons, new generations of weapons and new weapons.

Both the short-range and the long-range problems must be tackled: to institute immediately a most decisive shut-down measure, and to decide on the disarmament goal of what nuclear weapon forces should be tolerated to remain in the end if at all. The answer to the first query is a *banning of all nuclear weapon tests*, and to the second a *minimum nuclear weapons deterrent.*

The ban should really start with Research and Development work. But these are admittedly the activities most difficult to verify. An important recommendation for this domain is opening of all information that is accumulating and monitoring of the research under way, both subjects treated in connection with suggestions for an international disarmament verification agency.

The focal point for intervention must, in the form of internationally agreed banning, be concerned with the activities of testing the new weapons systems. Banning the testing of nuclear warheads has been our avowed aim for more than two decades. It was recognized as an urgent obligation when the Partial Test Ban Treaty was signed more than ten years ago. Now a comprehensive ban on testing of nuclear explosives must be taken up for serious and conclusive negotiations. Such testing is easily controlled, assertions to the opposite being excuses, void of factual content. Much more concerted attention must also be given to the issue of banning other tests. in the first instance testing of missiles. It ought, for instance, to be raised as a question whether the international community will continue to tolerate the closing off of large areas of the oceans for weapons testing, requesting ships of other nations to choose other routes. To me, and to many others, it appears to be a violation of 'the freedom of the seas'. Missile tests are also relatively easy to monitor with 'national means', particularly for nations which have at their disposal the whole technical spy apparatus with cameras and sensors of all kinds on aircraft and satellites.

Leaving the short-range but urgent questions of prohibiting weapon tests of various kinds, I am turning to the less discussed issue, namely what nuclear weapons should be permitted in the longer view. As the 'state of innocence' can never be re-created when once lost, there would seem to be scant hope for total nuclear disarmament. The clandestine production of even one bomb would then always be feared as a disaster out of the dark. Neither is it believable that there will ever be complete international control of all that pertains to the nuclear process: a tightly perfect Baruch system does not seem realizable. Greater safety should be feasible to attain through the existence of a few nuclear weapons for deterrence. How few, how constituted and how kept in custody is our most urgent task to examine. I would thus strongly recommend that both academic studies and negotiations return to the radical ideas of trying to fix rules for a minimum deterrent. It would seem as practically feasible as politically promising, if an international working group were established to study the parameters of such a nuclear shield, a minimum deterrent which through its size would be 'enough' and through its frozen stability of weapon properties would guarantee a stop qualitative competition.

Reversing the conventional arms race

So little thought has been given to measures that might be proposed in a serious attempt to block and, hopefully, reverse the race for building up the conventional armories. In the Geneva Disarmament Committee this issue does not even figure on the agenda. A systematic search must be made for measures which might be conducive to cope realistically with this problem. A first general approach is to lay bare the sombre facts. Active steps for introducing restraint must then be sought, either by the indirect method to reduce military expenditures across the board, or by the direct ones of prohibiting certain lines of production or of trading in conventional weapons.

(a) Disclosure of military expenditures

The first requirement for an international stocktaking of where the world is heading is to lay all the facts on the table. An annual yearbook on armaments and arms trade was published by the League of Nations until 1938. We should also remind ourselves that in the League's time the publication of national military budgets was recommended, a standardized accounting system was beginning to be developed and a number of states did, in fact, submit their military budgets to the League Secretariat in roughly standardized form.

There is no reason why these traditions should not be taken up again in the United Nations. More specifically, the military budgets should become part of public, international knowledge. There are difficult but not insurmountable technical problems to make figures for military expenditures comparable, particularly because some are often hidden in other parts of a national budget.

In the United Nations, proposals for more open divulgence of military costs have for a long time been a kind of Swedish specialization. More openness about military expenditures has always been, and should be advocated mainly in order to make possible international agreements on reducing these very expenditures. The simplest form of such agreements would aim at a straight reduction percentage-wise of the military budgets. But even just getting available standardized and fairly specialized national accounts for military expenditures according to categories of programs for procurement, opens up a way for critical comparisons and disarmament advocacy. Pointers will then be won for abolition or reduction of spending for specified weapons or other war preparations, for instance R&D expenditures.

(b) Disclosure of production and trade in arms

The first point to emphasize is that a prohibition against transfer of arms from outside states, presumed to be neutral, to any state taking part in warfare, is already laid down in the Hague Convention of 1907. Like many other stipulations of international law, this one has been habitually broken, blatantly and apparently thoughtlessly, and surprisingly without drawing much criticism either from legal scholars or from alert citizens. Neither have such violations, common as they are, been discussed within the United Nations. The arms flow to the warfaring states in the Middle East is only one example, though standing out above others because of the tremendous size of the deliveries of arms to adversaries, and its importance for making their wars more possible and more devastating.

Turning to the more general question in regard to the arms trade, the first request is full publicity about both arms trade and military aid. The aim should be a continual registration by a United Nations agency of all arms transfers whether taking place under one heading or another.

Proposals in the United Nations for more open divulgences of the arms trade—by Malta in 1965, United States in 1967 for sales to the Middle East, and by Denmark in 1958—have failed to win approval. They have met with resistance, mostly from underdeveloped countries which strongly argue that accounting for trade in arms, without accounting for production of arms, is discriminatory. Some of them resist having the knowledge revealed of how much of their foreign exchange reserves goes to paying for armaments. Countries which have no, or little, domestic production themselves often fear that any control over arms transfers, which they see as a continuation of their registration, would be directly discriminatory against them. The very fact that developed countries generally have most of their weapons produced at home, and even have a surplus to sell for profit or to give away, is felt as a discrimination against underdeveloped countries. My opinion is that they would raise less resistance to having their import of weapons registered if that were combined with a comprehensive registration also of all production of weapons in all countries.

(*c*) *Restraining production and trade in arms*

Registration of military budgets in some standardized form and publication of statistics on production and transference of weapons should be, and generally is, thought of as only a first step towards agreement on controls. Besides the ambition to reach agreed limitations of the whole budget, there are possibilities to seek out certain items in the budget, forbidding production and, of course, transference of certain specific weapons. Obvious candidates for such limited disarmament measures are in the first hand mass destruction weapons, such as biological and chemical ones which already are under a ban against use and close to a ban on production. Likewise, certain devices for military activities in the international ocean space belong to this category.

The very highest priority for a new campaign to reach an international prohibition must be given to the huge number of conventional weapons which are cruel and inhumane as, for instance, high-velocity weapons, fragmentation bombs, napalm, etc. These are now in ample stocks in many military establishments, and have even been used in wars, although that is contrary to international rules. Their eradication from national armories and the prohibition of their production should be thought of as pertaining to the re-establishment of the rules of international law.

The 'benign neglect' with which military violations of these rules have been treated must cease. To prohibit their transfer from one country to another of cruel weapons would be a suitable start.

Also, a new searchlight should be shed on the possibility of forbidding, or at least restraining, the production and transference of arms that are typically offensive in character, upholding as clear a demarcation line as ever possible in order to leave purely defensive capacities less restrained.

While these approaches address themselves to partial disarmament in regard to specific weapons in the category of what is euphemistically called conventional arms, there is a wider and more general approach, viz, attempts to bring to a halt the competition for ever more sophisticated weaponry. In regard to nuclear arms I stressed the need to stop the qualitative escalation and criticized the SALT negotiations for only focusing on quantities. Also for the broader field of conventional armaments the same request for stalling the qualitative arms race must be made, and stressed.

To some extent such a desirable development may be pursued through limitations of military budgets. Sheer cost considerations should in all countries support this request, as the constant modernization of weapons is so tremendously expensive, not least in regard to the drain upon scientists and technicians for R&D work.

Finally, new efforts must be made for not only worldwide but also for regional disarmament agreements. There is every reason why neighboring countries should voluntarily agree on mutual limitations of their armaments. The rest of the world has also every reason to be in favor of such regional agreements. That problem is first and foremost urgent on account of the possibility of nuclear weapon proliferation, but a movement thus begun could well lead to a joint interest in wider demilitarization.

*17. The law of war and dubious weapons**

The traditional principles of the law of war

War is an exceptional state of law in which destruction and killing are permitted, although not without restrictions. The relevant principles underlying the rules of traditional international law can be summarized as follows:

1. The prohibition of superfluous injury, hence
   (*a*) the prohibition of weapons which cause unnecessary suffering,
   (*b*) the prohibition of weapons which cause disproportionate suffering.

2. The distinction between civilians and soldiers, hence
   (*a*) the prohibition of an attack on civilians as such,
   (*b*) the duty to spare the civilian population as much as possible,
   (*c*) the prohibition of military acts which cause disproportionate suffering of civilians,
   (*d*) the prohibition of 'blind' or indiscriminate weapons,
   (*e*) the prohibition of weapons that are usually employed in such a manner that they cause disproportionate suffering or have indiscriminate effects.

3. The principle that the demands of humanity may prevail over the demands of warfare.

4. The principle that the demands of the conclusions of peace (including ceasefire and armistice) may prevail over the demands of warfare (prohibition of treachery).

* From *The Law of War and Dubious Weapons*, SIPRI, 1976.

Dubious weapons

Weapons have been used in World War II, in the Korean War (even by troops fighting under the banner of the UN), in Algeria and in Indo-China which can be considered 'dubious weapons' in the sense that they are morally repulsive and contrary to *traditional* principles, including the laws of humanity and the demands of the public conscience. The term 'dubious weapons' is used to denote all the modern weapons made possible by technology which may fall within categories forbidden by the laws of war. There are several categories of 'dubious weapons': (*a*) nuclear weapons, (*b*) biological and chemical weapons, (*c*) geophysical weapons, (*d*) incendiary weapons, including napalm, (*e*) small-calibre high-velocity weapons, (*f*) fragmentation weapons, and (*g*) delayed-action weapons (including booby traps). All these weapons have effects that may bring them under the rules concerning unnecessary suffering, disproportionate suffering, inhumane character, and indiscriminate effects. The question is how heavily these negative features weigh, and how far the military effectiveness or military necessity may be deemed to outweigh these negative features.

The point to be stressed here is that the standards of humanity play an independent role, apart from the element of unnecessary suffering. The principle of unnecessary suffering could easily be applied to the dum-dum bullet: the ordinary bullet was designed to disable the soldier, but the special construction of the dum-dum bullet superflously added to this 'being disabled' the unnecessary suffering of having a large wound which is more difficult to treat. This superfluous suffering could be removed, while still maintaining the capacity of the weapon to disable the soldier.

With many modern weapons, this sharp distinction between the capacity to disable and to inflict superfluous injury does not exist. The small-calibre high-velocity gun, according to the military, has qualities which other guns simply do not have: light in weight, and therefore easy to carry, but greater firepower due to its increased velocity. It cannot therefore be said that the suffering is unnecessary. It is perfectly possible to produce a light-weight gun with low velocity, as for example, the US 30-calibre carbine. The problem is that while civilians claim that this weapon is relatively humane, the military claim that it is not sufficiently powerful. Inhumane suffering is inevitable if the military effect is to be produced. However, the suffering might be considered to be *disproportionate* in comparison with the military gains.

Consequently, with respect to these weapons—and the same reasoning may apply to fragmentation weapons or incendiary weapons—the question is whether the *repulsiveness* of the effect of the weapon is so great that it outweighs the military advantages.

Progressive development of the law of war

The time has come to consider the question of whether—in view of technological developments in weaponry—new principles with respect to the laws of war concerning the prohibition of specific weapons should be added to the traditional ones.

In some cases doubt may exist about whether new principles must apply, or whether rules apply which logically derive from established principles with respect to newly developed weapons.

Such principles in line with already existing principles should, however, be expressly mentioned. They should be clearly and unequivocally recognized.

1. *The principle of proportionality* would lead to the following rules: (*a*) the prohibition of weapons which cause disproportionate suffering, and (*b*) the prohibition of military acts which cause disproportionate suffering.

2. *The aspect of the survival of mankind* should be taken into account with respect to forbidden weapons or forbidden activities. This would imply: (*a*) the principle that the survival of mankind prevails over any national interest, and (*b*) the principle that in the question of whether a specific weapon should be forbidden, not only should the inhuman character of the weapon be taken into account but also the danger for the integer continuation of the human race or groups thereof.

A fortiori, it seems that the survival of mankind should be recognized as a value which may lead to the prohibition of weapons or methods of warfare. In case such a value is recognized, the danger that mankind would be put in jeopardy should already be taken into account.

3. *The principle that the environment* should be taken into account with respect to forbidden weapons or forbidden activities would mean that the prohibition of specific weapons or specific action would also be based upon ecological considerations (the impact on nature, the destruction of its natural balance, the introduction of irrevocable processes).

4. *The principle of the threshold*, that the demands of humanity or of survival may imply the total prohibition of some kinds of weapons—notwithstanding the fact that a specific kind of use would not be contrary to the laws of war—because any use would mean a trespassing of a threshold, and would open up the road for an application of forbidden weapons by which the survival of mankind or the environment might be jeopardized.

This principle was applied at the time of the 1925 Geneva Protocol, when chemical weapons were totally prohibited. But it has never been expressly recognized. It would be advisable to do so, because some special cases could be mentioned in which the use of a specific kind of weapon would not be contrary to the principles of the law of war, but in which it would be necessary—if the states were to wish to ban that kind of weapon—to ban it totally and unconditionally. Experience in World War II, the Korean War and the Viet-nam War may have taught that mostly a specific kind of weapon, for example napalm, is employed in an illegal way. This may be a reason to forbid the weapon *in toto*.

In summary, all the principles of the laws of war relevant to dubious weapons may be presented as in the list below. These principles are valid today, and may all be considered as restatement, reformulation and adaptation of the traditional principles. They are pertinent to the question of whether the use of specific weapons should be outlawed.

1. Prohibited are weapons or munitions which *per se* cause unnecessary suffering to the disabled combatant.
2. The respect due to the civilian population leads to the prohibition of an attack on civilians, as such. Moreover, it leads to
   (*a*) the prohibition of weapons which cause incidental disproportionate suffering to the civilian population,
   (*b*) the prohibition of weapons which cannot be discriminate ('blind weapons'),
   (*c*) the prohibition of weapons which can be expected to be used in such a way that they cause disproportionate suffering to the civilian population or that they do not discriminate between civilians and combatants.
3. The cruel and repulsive character of weapons may lead to the conclusion that the laws of humanity and the demands of the public conscience should prevail over the favourable military aspects ('military necessity'), in such a way that they are prohibited notwithstanding their military usefulness.
4. Weapons may be prohibited because they can be expected to be generally used in such a way that the laws of humanity and the demands of the public conscience should prevail over the military usefulness.
5. The treacherous character of weapons may lead to their prohibition (because they would diminish the possibility of negotiations—cease-fire, peace negotiations).
6. Weapons which violate principles of the laws of war may be regarded as legitimate weapons because they are considered indispensable for the maintenance of peace.
7. Weapons may be prohibited because they threaten the integer existence of humanity or its constituent parts, peoples or civilizations ('survival value').
8. Weapons may be prohibited because they threaten to disturb the ecological balance by causing widespread, long-lasting or severe damage to the natural environment ('environment value').
9. Weapons may be prohibited because they can be expected to be generally used in such a way that they threaten the values of survival or environment.
10. Types of weapons may be prohibited totally—although a specific kind in specific circumstances would not violate the principles of the laws of war—because any use of this type of weapons would trespass a threshold between this type and other types of weapons, and thereby create the danger, through escalation, of the general use of these weapons.

Mankind should not be the slave of technology

The tendency exists to oppose prohibition of weapons if one's own military apparatus has a technological lead in this respect over other weapons systems. One does not easily give away such an advantage. This amounts to a partial rationality which concentrates on the power relation at a specific moment between specific opponents. But the history of weapon developments shows that it does not take long before other countries reach the same level of technological developments and obtain the same kind of weapons.

The question to be answered is whether military considerations would stand in the way of prohibiting weapons when both parties in an armed conflict have these weapons. If in such a situation the military advantages are balanced on both sides, humanitarian considerations may prevail. What is at stake here is not in the first place the prevalance of moral or humanitarian concepts, but the recognition that mankind should not be the slave of technology and should put a stop to the development of ever more sophisticated means of destruction. Ultimately, the primary issue is to prevent fighting that may put humanity itself in jeopardy. The survival issue looms behind all endeavours to prohibit repulsive and indiscriminate weapons.

The aim in the humanitarian law of warfare is to diminish the suffering in war. On this point the laws of war are more or less in competition with technology which produces ever more destructive weapons. The most crucial task of the law of armed conflicts will be to prohibit in the near future, before it is too late, the use of weapons of mass destruction, especially nuclear weapons.

Abbreviations and acronyms

ABM	Antiballistic Missile
ACDA	US Arms Control and Disarmament Agency
ADM	Atomic Demolition Munitions
ALCM	Air-Launched Cruise Missile
ASM	Air-to-Surface Missile
ASW	Anti-Submarine Warfare
ASSW	Anti-Strategic Submarine Warfare
AWACS	Airborne Warning and Control System
CB	Chemical and Biological
CBU	Cluster Bomb Unit
CBW	Chemical and Biological Warfare
CCD	Conference of the Committee on Disarmament
CEP	Circular Error Probability
COIN	Counterinsurgency Operations
CTB	Comprehensive Test Ban
ENDC	Eighteen-Nation Disarmament Conference
ERTS	Earth Resources Technology Satellite
EW	Electronic Warfare
FOBS	Fractional Orbital Bombardment System
GCD	General and Complete Disarmament
IAEA	International Atomic Energy Agency
ICBM	Inter-Continental Ballistic Missile
IRBM	Intermediate-Range Ballistic Missile
MARV	Manoeuvrable Re-entry Vehicle
MFR	Mutual Force Reduction Negotiations
MIMS	Multiple Independently Manoeuvrable Submunitions
MIRV	Multiple Independently targetable Re-entry Vehicle
MRBM	Medium-Range Ballistic Missile
MRV	Multiple Re-entry Vehicle
NASA	National Aeronautics and Space Administration (USA)
NATO	North Atlantic Treaty Organization
NPT	Treaty on the Non-Proliferation of Nuclear Weapons
PGM	Precision Guided Munitions
PNE	Peaceful Nuclear Explosion
PTBT	Partial Test Ban Treaty
R&D	Research and Development
RPV	Remotely Piloted Vehicle
RV	Re-entry Vehicle
SAC	Strategic Air Command

SALT	Strategic Arms Limitation Talks
SAM	Surface-to-Air Missile
SCAD	Subsonic Cruise Armed Decoy
SLBM	Submarine-Launched Ballistic Missile
SLCM	Submarine-Launched Cruise Missile
SRAM	Short-Range Attack Missile
SS	Diesel-powered attack submarine
SSB	Diesel-powered ballistic missile submarine
SSBN	Nuclear-powered ballistic missile submarine
SSG	Diesel-powered cruise missile submarine
SSGN	Nuclear-powered cruise missile submarine
SSM	Surface-to-Surface Missile
SSN	Nuclear-powered attack submarine
STOL	Short Take-Off and Landing
TOW	Tube-launched Optically tracked Wire-guided (anti-tank missile)
TTBT	Threshold Test-Ban Treaty
ULMS	Undersea Long-range Missile System
WTO	Warsaw Treaty Organization

Selected bibliography

ALEXANDER, A. S. *et al.*, *The Control of Chemical and Biological Weapons*, New York: Carnegie Endowment for International Peace, 1971.

ARON, R., *Peace and War*, New York: Praeger, 1967.

BAILEY, S. D., *Prohibitions and Restraints in War*, London: Oxford University Press, 1972.

BARNABY, C. F., and HUISKEN R. (eds.), *Arms Uncontrolled*, Cambridge, Mass.: Harvard University Press, 1975.

BEATON, L., *Must the Bomb Spread?*, Harmondsworth: Penguin, 1966.

BENOIT, E. (ed.), *Disarmament and World Economic Interdependence*, Oslo: Universitetsforlaget, 1967.

BOULDING, K. E. (ed.), *Peace and the War Industry*, Trans-action Books: Aldine Publishing Company, 1970.

BREDOW, W. von (ed.), *Economic and Social Aspects of Disarmament*, Oslo: BPP Publications, 1975.

BRENNAN, D. G. (ed.), *Arms Control, Disarmament and National Security*, New York: George Braziller, 1961.

BRODIE, B., *Strategy in the Missile Age*, Princeton: Princeton University Press, 1959.

BROWN, F. J., *Chemical Warfare: A Study in Restraints*, Princeton: Princeton University Press, 1968.

CARLTON, D. and SCHAERF, C., *The Dynamics of the Arms Race*, London: Croom Helm, 1975.

CLARKE, R. H., *We All Fall Down: The Prospect of Biological and Chemical Warfare*, London: Allen Lane, The Penguin Press, 1968.

CLARKE, R. H., *The Science of War and Peace*, London: Jonathan Cape, 1971.

DEUTSCH, K. W., *Arms Control and the Atlantic Alliance*, New York: Wiley, 1967.

DOUGHERTY, J. E., *How to Think About Arms Control and Disarmament*, New York: Crane, Russak, 1973.

DRIVER, C. P., *The Disarmers: A Study in Protest*, London: Hodder, 1964.

ENTHOVEN, A. and SMITH, W., *How Much is Enough? Shaping the Defense Program*, New York: Harper and Row, 1971.

FELD, L. T., GREENWOOD, T., RATHJENS, G. W. and WEINBERG, S. (eds.), *Impact of New Technologies on the Arms Race*, Cambridge, Mass.: MIT Press, 1971.

FRANK, L. A., *The Arms Trade in International Relations*, New York: Praeger, 1969.

GAILBRAITH, J. K., *How to Control the Military*, New York: Doubleday, 1969.

GLAGOLEV, I., *Why We Need Disarmament*, Moscow: Novosti Press.

HABERMAN, F. W., *Nobel Lectures: Peace*, Vol. I: 1901–1925, Vol. II: 1926–1950, Vol. III: 1951–1970, Amsterdam: Elsvier, 1972.

HARKAVY, R. E., *The Arms Trade and International Systems*, Cambridge: Ballinger, 1975.

HESS, H. W. (ed.), *Weather and Climate Modification*, New York: Wiley, 1974.

IKLE, F. C., *How Nations Negotiate*, New York: Praeger, 1964.

JUNGK, R., *Brighter than 1000 Suns*, London: Golancz, 1958.

KISSINGER, H. A., *Nuclear Weapons and Foreign Policy*, New York: Doubleday, 1958.

LAPP, R. E., *Kill and Overkill: The Strategy of Annihilation*, London: Weidenfeld, 1963.

LAURIE, P., *Beneath the City Streets: A Private Enquiry Into the Nuclear Preoccupations of Government*, London: Allen Lane, Penguin, 1970.

LONG, F., and RATHJENS, G. W. (eds), *Arms, Defense Policy, and Arms Control*, New York: Nortoh, 1976.

MARKS, A. W. (ed.), *NPT: Paradoxes and Problems*, Washington D.C.: Carnegie Endowment for International Peace, 1975.

MARTIN, L. W., *The Sea in Modern Strategy*, London: Chatto & Windus, 1967.

MCDONALD, J. F., *Weather Modification*, US Senate Committee on Foreign Relations. Hearing before the Subcommittee On Oceans and International Environment, Washington, D.C.: US Government Printing Office, 1974.

MCGRATH, T. D., *New Perspectives on Anti-Submarine Warfare and Oceanology*, Washington, D.C.: Data Publications, 1967.

MELMAN, S., *The Permament War Economy: American Capitalism in Decline*, New York: Simon & Schuster, 1974.

NATIONAL ACADEMY OF SCIENCES, *Weather & Climate Modification, Problems and Progress*, Washington, 1973.

NOEL-BAKER, P., *The Arms Race: A Programme For World Disarmament*, London: Calder, 1958.

RICHARDSON, L. F., *Statistics of Deadly Quarrels*, Chicago: Quadrangle Press, 1961.

ROBINSON, J. P., *CBW: An Introduction and Bibliography*, Los Angeles: Political Issue Series, Vol. 3, No. 2, 1974.

RUSSET, B. M. (ed.), *Peace, War, and Numbers*, London: Sage Publication, 1972.

SARKESIAN, S. (ed.), *The Military-Industrial Complex: A Reassessment*, London: Beverly Hills, 1972.

SCHILLING, W. R., HAMMOND, P. Y., SNYDER, G. H., *Strategy, Politics and Defense Budgets*, New York, 1962.

SCOVILLE, H., and OSBORN, R., *Missile Madness*, Boston: Houghton Mifflin, 1970.

SINGER, D. J., *Deterrence, Arms Control and Disarmament*, Ohio State University Press, 1962.

SOKOLOVSKY, M. V. D., *Military Strategy: Soviet Doctrine and Concepts*, New York: Praeger, 1963.

STONE, J. J., *Containing the Arms Race*, Cambridge, Mass.: MIT Press, 1966.

THOMAS, A. V. W. and THOMAS, A. J., *Legal Limits On the Use of Chemical and Biological Weapons*, Dalles: Southern Methodist University Press, 1970.

UNITED NATIONS, *Basic Problems of Disarmament*. Reports of the Secretary-General, New York: United Nations, 1970.

UNITED NATIONS, *The United Nations and Disarmament*, 1945–1970, New York: United Nations, 1970.

UNITED NATIONS, *Disarmament and Development. Report of the Group of Experts on the Economic and Social Consequences of Disarmament*, New York: United Nations, 1972.

UNITED NATIONS, *Napalm and Other Incendiary Weapons and All Aspects of Their Possible Use*, Report of the Secretary-General, New York: United Nations, 1973.

UNITED NATIONS, *Reduction of the Military Budgets of the States Permanent Members of the Security Council by 10 per cent and Utilization of Part of the Funds Thus Saved to Provide Assistance to Developing Countries.* Report of the Group of UN Consultant Experts, New York: United Nations, 1974.

US ARMS CONTROL AND DISARMAMENT AGENCY, *World Military Expenditures and Arms Trade 1963–1973*, Washington D.C.: ACDA, 1975.

US ARMS CONTROL AND DISARMAMENT AGENCY, *Selected Background Documents Relating to Mutual and Balanced Force Reductions*, Part I. Disarmament Document Series ref. no. 611, May 17, 1973. Part II. DDS ref. no. 619, November, 1973. Part III. DDS ref. no. 637, May 5, 1975 a.

US CONGRESS, HOUSE OF REPRESENTATIVES, COMMITTEE ON FOREIGN AFFAIRS, SUBCOMMITTEE ON NATIONAL SECURITY POLICY AND SCIENTIFIC DEVELOPMENTS, *Chemical-Biological Warfare: U.S. Policies and International Effects*, Washington, D.C.: US Government Printing Office, 1970.

US CONGRESS, SENATE, COMMITTEE ON FOREIGN RELATIONS, *The Geneva Protocol of 1925*, Washington, D.C.: US Government Printing Press, 1972.

US CONGRESS, HOUSE OF REPRESENTATIVES, COMMITTEE ON FOREIGN AFFAIRS, SUBCOMMITTEE ON NATIONAL SECURITY POLICY AND SCIENTIFIC DEVELOPMENTS, *U.S. Chemical Warfare Policy*, Washington, D.C.: US Government Printing Office, 1974.

US CONGRESS, SENATE COMMITTEE ON FOREIGN RELATIONS, *Weather Modification*. Hearings before the Subcommittee on Oceans and International Environment, Washington, D.C.: US Government Printing Office, 1974.

US CONGRESS, HOUSE OF REPRESENTATIVES, COMMITTEE ON APPROPRIATION SUBCOMMITTEES ON THE DEPARTMENT OF DEFENSE. Hearings on Binary Chemical Munitions, 9 June 1975, *Department of Defense Appropriations for 1976*, Washington, D.C.: US Government Printing Office, 1975 (Part 9, pp. 206–262).

US Congress, Senate, Committee on Foreign Relations, *Détente*, Hearings in August, September and October 1974, Washington, D.C.: US Government Printing Office, 1975.

US Congress, Senate, Committees on Armed Services, *United States Soviet Military Balance*, A Study by the Library of the Congress, Washington, D.C.: US Government Printing Office, 1976.

Weidenbaum, M. L., *The Economics of Peacetime Defense*, New York: Praeger, 1974.

Wolfe, Th. W., *Soviet Power and Europe* 1945–1970, Baltimore: The John Hopkins Press, 1970.

World Health Organization, *Health Aspects of Chemical and Biological Weapons*. Report of a WHO Group of Consultants, Geneva: WHO, 1970.

Wright, Q., *A Study of War*, Chicago: University of Chicago Press, 1971.

Yarmolinsky, A., *The Military Establishment*, New York: Harper & Row, 1971.

York, H. F. (ed.), *Arms Control*, Readings from the *Scientific American*, San Francisco: Freeman, 1973.

York, H. F., *Race to Oblivion: A Participants View of the Arms Race*, New York: Simon and Schuster, 1970.

Young, E., *A Farewell to Arms Control?*, Harmondsworth: Penguin, 1973.

Index

U

V

W

Y

Z

SIPRI Books

SIPRI Yearbook of World Armaments and Disarmament 1968/69. 1969. 440 pp. 93 tables, charts and maps. Cloth bound (ISBN 91-85114-03-0). Paperback. (ISBN 91-85114-04-9).

SIPRI Yearbook of World Armaments and Disarmament 1969/70. 1970. 540 pp. 91 tables and charts. Cloth bound. (ISBN 91-85114-07-3). Paperback (ISBN 91-85114-06-1).

World Armaments and Disarmament, SIPRI Yearbook 1972. 1972. 600 pp. 116 tables and charts, 7 maps. Cloth bound. (ISBN 91-85114-12-X).

World Armament and Disarmament, SIPRI Yearbook 1973. 1973. 515 pp. 68 tables, 13 charts, 2 photographs. Cloth bound. (ISBN 91-85114-19-7).

World Armaments and Disarmament, SIPRI Yearbook 1974. 1974. 520 pp. 83 tables and charts. Cloth bound. (ISBN 0-262-19129-6).

World Armaments and Disarmament, SIPRI Yearbook 1975. 1975. 618 pp. 115 tables and charts. Clothbound. (ISBN 0-262-19140-7).

World Armaments and Disarmament, SIPRI Yearbook 1976. 1976. 493 pp. 82 tables and charts. Clothbound. (ISBN 0-262-19149-0).

The Arms Trade with the Third World. 1971. 910 pp. 141 tables, 41 charts, 6 maps. Clothbound. (ISBN 91-85114-09-X).

The Problem of Chemical and Biological Warfare. 6 volumes. 1971–1975. Clothbound. (List of titles available on request).

Chemical Disarmament: Some Problems of Verification. 1973. 184 pp. 52 tables, 11 charts, 5 maps. Paperback. (ISBN 91-85114-20-0).

Force Reductions in Europe. 1974. 106 pp. 12 tables. Paperback. (ISBN 91-85114-21-9).

Nuclear Proliferation Problems. 1974. 312 pp. 36 tables, 7 charts, 1 map. Clothbound. (ISBN 0-262-10015-0).

Oil and Security. 1974. 197 pp. 11 tables, 1 chart. Paperback. (ISBN 91-85114-25-1).

Tactical and Strategic Antisubmarine Warfare. 1974. 148 pp. 8 tables. Clothbound (ISBN 0-262-20031-7).

Arms Trade Registers: The Arms Trade with the Third World. 1975. 176 pp. 21 tables, 8 charts, 2 figures. Clothbound. (ISBN 0-262-19138-5).

The Nuclear Age. 1975. 148 pp. 37 tables, 8 figures, glossary. Clothbound. (ISBN 0-262-19136-9).

Safeguards Against Nuclear Proliferation. 1975. 114 pp. Clothbound. (ISBN 0-262-19137-7).

Chemical Disarmament: New Weapons for Old. 1975. 151 pp. 10 tables, 5 figures. Paperback. (ISBN 91-85114-27-8).

Delayed Toxic Effects of Chemical Warfare Agents. 1975. 60 pp. 3 tables, 5 figures. Paperback. (ISBN 91-85114-29-4).

Incendiary Weapons. 1975. 255 pp. 30 tables, 5 figures. Clothbound. (ISBN 0-262-19139-3).

Medical Protection against Chemical-Warfare Agents. 1976. 166 pp. 47 tables, 27 figures, 1 plate. Paperback. (ISBN 91-22000-44-5).

The Law of War and Dubious Weapons. 1976. 78 pp. Paperback. (ISBN 91-85114-31-6).

Publishers

SIPRI books are published by collaboration between SIPRI and three publishers. Other publications may be obtained from SIPRI.

In the United States

HUM Humanities Press Inc.
171 First Avenue
Atlantic Highlands, New Jersey 07716

In the western hemisphere and in Britain and the Commonwealth

MIT The MIT Press
28 Carleton Street
Cambridge, Mass. 02142, USA
and
126 Buckingham Palace Road
London SW1W 9SD, England

In the rest of the world

A & W Almqvist & Wiksell International
P.O. Box 62
S-101 20 Stockholm 1, Sweden